Science for the Curious Photographer

An Introduction to the Science of Photography

Science for the Curious Photographer

An Introduction to the Science of Photography

Charles S. Johnson, Jr.

A K Peters, Ltd.
Natick, Massachusetts

Editorial, Sales, and Customer Service Office
A K Peters, Ltd.
5 Commonwealth Road
Natick, MA 01760
www.akpeters.com

The following products and trademarks are used throughout the book: AT&T Bell Labs, B+W, Canon, Cibachrome, Ciga-Geigy Corp., CoCam?, Cokin, ColorMunki, CombineZM (freeware), Dynamic-Photo HDR, Eastman Kodak, Kenko, Kodak, Kodachrome, Epson, Fairchild Semiconductor, Focus Magic, FocalBlade, Foveon, Fujichrome Provia, Fujifilm, Helicon Focus, Heliopan, Hoya, Ilfochrome, Intel, Kirk, Life Pixel, Leica, MaxMax, Manfrotto, Nikon, Olympus, Photokit Sharpener, Photomatix, PhotoPro RGB, Photoshop, Polaroid Corp., Rolleicord, Schott, Schneider-Kreuznach, Sigma, Sony, Spyper3, Tamron, Texas Instruments, Tiffin, Wratten, and Zeiss.

Library of Congress Cataloging-in-Publication Data

Johnson, Charles S. (Charles Sidney), 1936–
Science for the curious photographer : an introduction to the science of photography / Charles S. Johnson, Jr.
p. cm.
Includes bibliographical references and index.
ISBN 978-1-56881-581-7 (alk. paper)
1. Photography. I. Title.
TR146.J556 2010

770—dc22

2009047521

Printed in India
14 13 12 11 10

10 9 8 7 6 5 4 3 2 1

To Ellen, my wife and best friend

Table of Contents

Preface

My love of photography started very early. As a teenager I worked in a small full-service photography shop. Portraits were made, events were photographed, snapshots were developed and printed, and equipment was sold. From that experience, I learned about photographic techniques and the value of quality cameras and lenses. I started developing and enlarging my own photographs, and I searched for ways to learn more. Fortunately, I found the book *Lenses in Photography* by Rudolf Kingslake in the photography shop, and I studied it diligently. I still have that book and refer to it frequently. At that time (the 1950s), my uncle was serving with the Air Force in Germany, and he was able to buy fine cameras for me. First, I got a Zeiss Ikonta 35. It was bare bones, with no rangefinder or light meter, but it had a wonderful Zeiss Tessar *f*/2.8 lens. It was great for documenting sports and other high school events. Later, when I started doing freelance photography (while still in high school), I ordered a Rolleicord III medium format camera. It served me well and is still functional.

From the beginning I was fascinated by all aspects of photography. I love the equipment, the techniques, the processing of images, and, of course, the chance to photograph interesting things. Photography also provided a summer job and a doorway to business and social interactions. Photographs documenting those years reveal small-town life in the 1950s and a few cheesecake pictures as well. My career as a scientist and a university professor took me away from photography for many years, but in the past decade I have returned to that early love. I spend a lot of time on nature photography now, and I enjoy photo shoots with the Carolinas' Nature Photographers Association (CNPA). Of course, everything is digital, so the chemical darkroom is no longer necessary. This has given me a new world of opportunities and an array of new subjects to understand.

In my case, understanding the way photography works has increased enjoyment of it. Each new question is a challenge. The process of working through the concepts of photography from basic optics and image sensors to human perception of color and the appreciation of beauty was an exhilarating experience for me. I have written this book for those who also enjoy photography, and who want to know more about their photographic equipment and the operation of their eyes and brain as well. The book is specifically aimed at those who enjoy science and are not afraid of a little math. Of course, perfectly good photographs can be made by those who have no interest in the scientific side of photography. They see a clean separation between the scientific part and the artistic part, and they reject the scientific part. On the other hand, some great photographers and other artists as well have benefited from a knowledge of their media and ways to get the most out of it: Ansel Adams comes to mind. In addition to making awe-inspiring photographs, Adams wrote technical books on cameras, negatives, and prints. To each his/her own, but I believe that in photography, as elsewhere, knowledge is power.

I have worked on this book for four years, trying the patience of my wife and friends. I appreciate comments from all those who have read sections of it at various stages of its gestation. I am sure to leave out some generous and helpful people, but here is at least a partial list of those who have contributed at various times with corrections and advice: John Fowler, Archibald Fripp, Richard Jarnagin, and Calvin Wong.

—CHARLES S. JOHNSON, JR.

CHAPTER 1

What Is Photography?

The painter constructs,
the photographer discloses.
—SUSAN SONTAG

You don't take a photograph,
you make it.
—ANSEL ADAMS

There are two sides to photography. First, photography is the capture and display of images by means of film or an electronic sensor, and, second, photography is the art of taking and presenting photographs. As commonly practiced, photography is inseparable from cameras. Of course, *photography* means "writing with light" and *writing* is really the operative word. When photography was invented in 1839, the thing that was discovered was the means for permanently capturing images. Cameras of various kinds had, in fact, been available for centuries.

The original camera, known as the *camera obscura* (see Figure 1.1), was nothing more than a dark room with a small hole (aperture) in one wall and an inverted image on the opposite wall created by light rays passing through the aperture. The wonderful image-forming property of a small aperture was noted by the philosophers Mo-Ti in China and Aristotle in Greece in the 5th and 4th centuries BCE, respectively, although apparently without an understanding of how it was accomplished. It is clear that scientists in the Western world from at least the time of Leonardo da Vinci (*c.* 1490) were aware of the camera obscura, and at some point it was discovered that the image

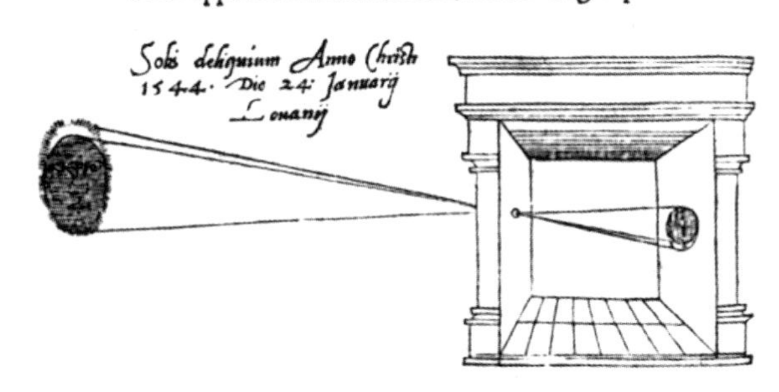
illum in tabula per radios Solis, quàm in cœlo contingit: hoc eſt, ſi in cœlo ſuperior pars deliquiũ patiatur, in radiis apparebit inferior deficere, vt ratio exigit optica.

Sic nos exactè Anno .1544. Louanii eclipſim Solis obſeruauimus, inuenimusq; deficere paulò plus q̃ dex-

FIGURE 1.1. The camera obscura was used by Reinerus Gemma-Frisius in 1544 to observe an eclipse of the sun.

FIGURE 1.2. Louis Mandé Daguerre (left) and Henry Fox Talbot (right).

quality and intensity could be improved by enlarging the aperture and inserting a convex lens of the appropriate focal length. The portable camera obscura, a box with a lens on one side and some means of viewing the image, became popular with artists as an aid in representing perspective in paintings. For example, the 16th-century Dutch painter Johannes Vermeer (1632–1675) almost certainly used a camera obscura to see the correct representation of perspective for his paintings. By the 19th century these devices were essentially box cameras without photographic film.

In the early 19th century, many individuals were experimenting with sensitized materials that darkened when exposed to light and produced fleeting images, so proper credit for the invention of photography is diffuse and controversial. Photography as we know it dates from 1839 when two men independently reported processes for capturing images in the camera obscura. Their disclosures initiated explosive developments in image-making around the world. The Frenchman Louis Mandé Daguerre discovered a method for producing a permanent image on a silver surface; while, in England, Henry Fox Talbot created permanent images on paper treated with a mixture containing silver chloride. In Daguerre's images, the areas exposed to light and properly processed were highly reflecting; and, therefore, there was a natural (positive) appearance though, of course, without color (monochrome). These images, which were called daguerreotypes, unfortunately, could not easily be reproduced. The striking images obtained by Daguerre were an instant hit, however, and most contemporaries considered him to be the inventor of photography. (See Figure 1.2.)

In marked contrast, Talbot's images were initially unpleasing because the (bright) exposed areas were found to be dark on the paper. In other words, a negative image was produced. That turned out to be a great advantage, however, because the negative could be combined with another sheet of sensitized paper and exposed to light to produce a positive copy, and that procedure could be repeated to produce multiple copies. Of course, paper is translucent rather than transparent, and it was not until the 1850s that transparent negatives could be obtained. The terms *photography* and *photograph* are usually attributed to Sir John Herschel, who included them in a paper that he read to the Royal Society of London in 1839. Herschel also deserves credit for advancing photographic science by discovering how to stabilize silver images; however, credit for the name *photography* is controversial. The term photography may actually have been introduced earlier by an artist named Hercules Florence working in Brazil in 1833. Florence, who used sensitized paper to copy drawings, did not report his work, and as a consequence, he had little influence on the development of photography.

For the next 160 years, silver-sensitized paper and film coupled with the negative/positive process, dominated photography; and it was only after the year 2000 that photoelectric detectors and powerful, yet inexpensive, computers challenged film-based photography. Replacing film with sensors and computer memory has not yet basically changed photography; however, computer manipulation of images has turned out to be a revolutionary development. Even images captured on film are now routinely scanned into computers and digitized so that they are also subject to modification. If computer image manipulations were limited to the types of things that photographers were already doing in the dark room to correct exposure, hold back or burn in areas, change contrast, etc., there would be no fundamental change in photography. But now the changes can be so extensive and subtle that the boundaries of photography are continuously being tested.

It has been said that photographers reveal while artists create. Software for manipulating photographic images and even creating realistic images from scratch is fundamentally changing this equation. Illustrators using computer graphics have almost unlimited ability to produce realistic images of any type. Photojournalists, on the other hand, must have their creative inclinations severely limited by a code of professional ethics and perhaps by authentication software that can spot even microscopic changes in digital images.

Anyone who has viewed recent movies knows that amazing things can be done to produce realistic images of things that never existed. The opening scenes in *Day After Tomorrow* show a flight over ocean, ice, and cliffs in Antarctica. It is beautiful and impressive. How was it done? A helicopter flight over those remote areas would be costly and dangerous, so the producer decided to create the scenes entirely with computer graphics. And what about the magnificent scenes in the epic *Troy*? Does anyone believe that 1000 or even 100 ships were constructed, or that 75,000 Greek warriors took part in the battle? We can all enjoy the endless possibilities for image-making, but we can no longer (if ever we did) believe in what the images show.

So, ultimately, what is photography? Does it matter that wrinkles can be removed from faces and heads can be switched? Do we care if it is easy to move an alligator from a zoo to a natural area or a hummingbird from a feeder to a flower? These tricks are still rare enough that gullible observers may marvel at how such "difficult" photographs could be obtained. I think we are seeing the emergence of a new art form, but I am not sure where that leaves photography. Will "pure" photography remain when everyone has an incentive to improve the images they obtain and it is so easy to do? Is there any merit in maintaining photography with minimum manipulation for recording the world as it is? The future will tell. In fact, "truth" is found in some novels and paintings and in some photographs, but it must be tested and verified by wise readers and observers.

Questions raised by the concepts of reality and truth in visual images are much more complex than may be thought. Later in the book I discuss the operation of the human visual system and its relation to our awareness of the world. It is fair to say that our eyes and brain create the illusion of a full-color, three-dimensional world. It is an illusion, because the images projected on the retinas of our eyes do not provide enough information for the construction of a unique worldview. The brain fills in details based on a sort of automatic inference system that is influenced by both the evolution of the human brain and the experience of the individual. The result is that we see, at first glance, petty much what we expect to see. One should also realize that digital cameras basically compute pictures from captured light. The computation is not straightforward, and there is a lot of room for "enhancement" of the final image. The new field of computational photography is influencing the images produced by our cameras and the special effects we see in movies. It is an interesting time to be alive and maybe a little disturbing as well.

Further Reading

N. Rosenblum. *World History of Photography, Third Edition.* New York: Abbeville Press, 1997. (This is a tour of photography from 1839 through the film era, including both art and the technical details.)

M. R. Peres (Editor). *Focal Encyclopedia of Photography, Fourth Edition*. Amsterdam: Elsevier, 2007. (Although it is uneven and already somewhat dated, this text provides extensive coverage of theory, applications, and science.)

CHAPTER 2

What Is Light?

The study of light has resulted in achievements of insight, imagination, and ingenuity unsurpassed in any field of mental activity; it illustrates, too, better than any other branch of physics, the vicissitudes of theories.

—SIR J.J. THOMPSON

All the fifty years of conscious brooding have brought me no closer to the answer to the question, "What are light quanta?" Of course today every rascal thinks he knows the answer, but he is deluding himself.

—ALBERT EINSTEIN

We can wax eloquent: Light is universally revered. It represents illumination and understanding. It represents our greatest treasure, as in the "light of our lives." We believe that bad things happen under cover of darkness, and evil works cannot stand the light of day. Light sets the pattern of our days. When light goes away, we sleep. Photography involves writing with light, and it uses the information in light to capture moments in our lives so that they are preserved and can be replayed. In this way photography brings light to our past.

But here we need useful rather than poetic definitions that can help us understand photography. We all know the difference between light and dark, but what is light? The problem is that most definitions introduce other words that need to be defined. I will start with several ideas that are correct but not complete as a step toward understanding what we mean by *light*. First, light is a form of energy that comes to us from some source and can make our environment visible to us. Of course, even with our eyes closed we can tell the difference between bright light and no light. Another useful statement is that light is a kind of radiation that is visible to us. Again this implies energy, the ability to do work and affect change.

All of our energy, other than that associated with radioactivity, comes to us directly or indirectly from the sun, and the sun is our major source of light. The sun provides bright white light in the middle of the day and less intense yellow and red shades

FIGURE 2.1. The sun, primary source of light and energy.

in the early morning and late afternoon, at least as seen from the surface of the Earth (Figure 2.1). This sentence brings up the concepts of intensity and color, both of which are related to human perceptions. Therefore, the study of light in photography requires not only precise mathematical definitions of brightness and color, but also methods of quantifying perceptions of brightness and color. The latter gets us into the realm of psychophysics.

Later chapters will be devoted to the measurement of light, the manipulation of light, and the recording of light. Here I consider only the primary physical properties of light, starting with a bit of history. In the 4th and 5th centuries BCE, the ancient Greeks speculated extensively about the nature of light. Some (Empedocles, Euclid, and Plato) thought that light is projected from the eye. Lucretius believed that particles of light are sent out from the sun, while Pythagoras held that light particles are emitted by objects. Aristotle even offered a theory of wave propagation of light. Some of these ideas have elements of truth, but they were not particularly useful and did not advance knowledge because they were not based on a consistent theoretical framework and were not tested. In fact, these ideas only provided a set of possibilities.

There were, of course, some observations and insights in the antique world that carry over into the scientific study of light. For example Euclid (4th and 3rd centuries BCE) noted that light appears to travel in a straight line, and Hero (1st century BCE) concluded from the study of reflections that light follows the shortest path between two points. However, modern ideas about the nature of light and optics have their origin in the 17th century. At that time, two ideas dominated theory and experiment: first, that light is a propagating wave and, second, that light consists of streams of particles (corpuscles).

One of the most influential scientists of that period was Sir Isaac Newton (1642–1727, Figure 2.2), who had carried out exquisite optical experiments to determine the nature of color; and who developed a theory of color we still use. He concluded that light rays consist of particles rather than waves because light travels in a straight line. Anyone who has played with a laser pointer might reach the same conclusion. Now we know that this view is incomplete because, among other things, shadows do not have sharp edges, and there is much evidence of diffraction in the everyday world. I will treat diffraction in detail later; here it suffices to say that diffraction refers to the deflection and decomposition that occurs when light interacts with sharp edges of opaque objects, narrow slits, or collections of narrow slits (gratings). In fact, in 1663 James Gregory noticed that sunlight passing through a feather is diffracted into spots of different colors and he wrote, "I would gladly hear Mr. Newton's thoughts of it." Francesco Grimaldi even published a book in 1665 on the effects of small apertures on the passage of light beams that seemed to indicate wave-like properties for light. Apparently, Newton did not consider the possibility that light might be a wave with a very small wavelength so that it would indeed appear to travel in a straight line unless one looked very closely.

FIGURE 2.2. Sir Isaac Newton (1642–1727), who dispersed light with a prism and developed a theory of color.

The particle theory was, in fact, undermined by continuing observations of diffraction and the presentation of a wave theory of light by Christiaan Huygens in 1678 that explained such phenomena, albeit with certain unproved assumptions. According to the wave theory, diffracted intensity patterns, which consist of bright and dark lines, result from interference effects—that is to say, the addition of the amplitudes of waves. In this way, the positive and negative peaks can reinforce to give intensity maxima or cancel to produce the dark places. This process is shown in Figure 2.3.

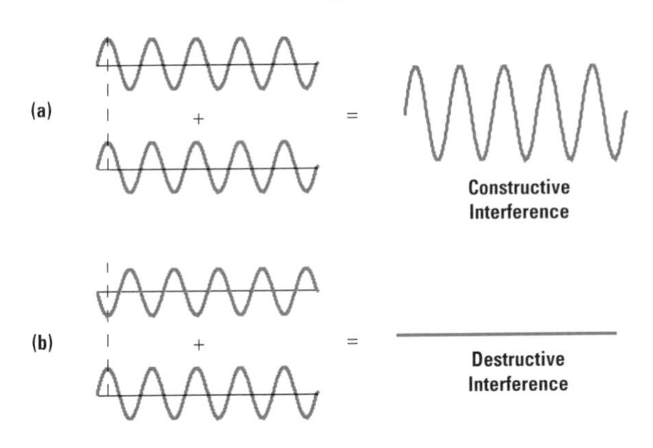

FIGURE 2.3. The (a) constructive and (b) destructive interference of waves.

The validity of the wave theory was confirmed by a number of scientists. Most notably, Thomas Young (1802), who demonstrated the interference (diffraction) of light by passing a light beam through two closely spaced slits in an opaque sheet and displaying the resulting pattern of bright and dark lines. Also, Joseph Fraunhofer and Augustine Fresnel investigated diffraction effects using small holes of different shapes and presented rigorous theories for calculating diffraction patterns, thus advancing the wave hypothesis. The diffraction of light from a compact disc (Figure 2.4) where there are 625 tracks per millimeter provides a beautiful example.

By the early 19th century, the wave properties of light were well known, and there was a practical understanding of the way light rays behave and of the basis of perceived color resulting from the mixing of light beams. At that time, however, there was still the open question, "What is light?"

FIGURE 2.4. Diffraction of light from the surface of a compact disc. The pitch of the data tracks is 1.6 μm, and each angle of reflection selects a specific spectral color.

One line of evidence was provided by the speed of light. To casual observers, the speed of light appears to be infinite, but well-planned experiments show that is not the case. An early, approximate value of the speed of light had been determined from observations of eclipses of the moons of Jupiter by Olaf Romer (1676), and by 1860 there were measurements of the speed of light in air from the study of light pulses (Armand Fizeau) and rotating mirrors (J. L. Foucault). The conclusion of this part of the story was provided by James Clerk Maxwell's monumental achievements in the 1860s. Maxwell (see Figure 2.5) developed a set of partial differential equations to describe electric and magnetic fields. He concluded that oscillating electric fields are always accompanied by oscillating magnetic fields and together they propagate through space as electromagnetic waves. In 1862 he computed the velocity of electromagnetic waves and discovered

FIGURE 2.5. James Clerk Maxwell (1831–1879), who identified light as electromagnetic radiation.

that they have the same velocity as light. This revelation prompted him to write, in an 1864 paper, "The velocity is so nearly that of light that it seems we have strong reason to conclude that light itself (including radiant heat and other radiations) is an electromagnetic disturbance in the form of waves propagated through the electromagnetic field according to electromagnetic laws." So the simple answer to the question "What is light?" is that light is electromagnetic (EM) radiation.

But that is not the end of the story. In 1905 Albert Einstein (see Figure 2.6) showed that light is composed of quanta or particles of energy, and every experiment since that time confirms his conclusion. What he had discovered was that light quanta are necessary to explain the interaction of light with electrons in metals. For this amazing conceptual breakthrough Einstein was awarded the Nobel Prize in 1921. Newton was right, but for the wrong reasons. We always detect light quanta, now called *photons*, but the wave theory lets us compute the probability of finding a photon at a certain location. Each photon carries an amount of energy that depends on the frequency of the light, and the number of photons arriving each second determines the intensity. A few photons are required to activate a nerve in the eye, and the creation of a latent image in photographic film requires the absorption of photons by atoms in silver halide grains. By the late 20th century, instruments for counting photons were widely available.

So we have a beautiful and subtle story: Light is made of particles, and light is a wave. Maxwell's equations are sufficient for describing light when there are many photons and the quantum theory is necessary when there are only a few. It is no wonder that earlier scientists had such a hard time getting it straight.

FIGURE 2.6. Albert Einstein (1879–1955), who discovered that light is quantized.

In later chapters we will explore in detail how the nature of light is manifest in photography and especially digital photography. We will see how the wave nature of light ultimately determines the way lenses work but also limits the resolution of all lenses. For example, the size of the image of a distant star is determined by diffraction, a wave phenomenon, as well as by imperfections in the lens and the resolving power of film or an electronic sensor. Also, detection and recording of an image depends on the collection of photons in each picture element (pixel) of the detector or by grains of silver halide in a photographic film. When the light level is low, the number of photons in each pixel may be small, so that deviations from the average will be significant. Therefore, different pixels will receive different numbers of photons and the image will appear to be "noisy." We shall see that the laws of physics ultimately limit the quality of images and force us to make compromises in camera and lens designs. I expand on this topic and define the signal-to-noise ratio in Chapter 16.

CHAPTER 3

The Camera—An Introduction

The camera is an instrument
that teaches people how to see
without a camera.

—DOROTHEA LANGE

New cameras don't just capture
photons; they compute pictures.

—BRIAN HAYES

3.1 Introduction

Much of this book is concerned with the components of photographic cameras and the way they operate. This chapter outlines the major features of cameras as an introduction to the important topics that will be treated in more detail in later chapters. Also, I will take this opportunity to define the parameters that characterize lenses, films, and sensors. In later chapters I consider the resolving power of lenses, the sensitivity of detectors, the effects of diffraction, etc. An important goal of this part of the study is to show under what conditions an "equivalent image" can be obtained with cameras having film/sensors of different sizes.

A camera is essentially a box (light-proof housing) that contains a lens and a light sensor. We also expect the camera to include an aperture to control the amount of light passing through the lens, a shutter to control the duration of the exposure, and some kind of viewfinder to frame the object to be photographed. The earliest cameras were quite large, and the light detector was a sensitized photographic plate. The plate was exposed and then developed to make a negative; and the negative was used to make a positive contact print of exactly the same size. Since there was usually no enlargement, the lenses were not severely tested and the resulting prints often appeared to be of good quality. Of course, some photographic processes, e.g., daguerreotype and tintype, directly produced positive images on metal plates. (It is interesting to note that these images were reversed left to right so that, for example, in a famous Civil War portrait of President Lincoln his mole appeared on the wrong cheek. This is not a problem with prints made from negatives, since they are reversed twice.)

3.2 The Camera Body

In the first few decades after the invention of photography, the way to make a larger photograph was to use a larger camera. In spite of efforts to develop enlargers for photographs, many professional photographers preferred large plate film cameras. It is reported that William Henry Jackson hauled a 20 in. × 24 in. glass-plate camera on mule-back to photograph the western mountains. Also, Ansel Adams (1902–1984) often used a camera with 8 in. × 10 in. film plates. But certainly George R. Lawrence takes the cake. In 1901 he photographed a train with a camera that could take photographs measuring 4 ft. 8 in. × 8 ft. That gives a whole new meaning to the term "large-format photography." In contrast to this, the public clamored for portable cameras with small negatives that could be enlarged, an early example of which is shown in Figure 3.1. (See Appendix A for an historical note on enlargers.)

FIGURE 3.1. A Rochester Camera Co. PREMO SR 5 in. × 7 in. portable camera (1896). *(Camera from the collection of Denise Tolley.)*

Historically, cameras were manual devices; the photographer was expected to load the film, set the lens aperture, set the shutter speed, adjust the lens to obtain the desired focus, and even cock the shutter before making the exposure. Larger, professional cameras often had film holders for sheet film while smaller cameras that were available later used roll film that could be cranked to move fresh film into position. Also, professional cameras usually had interchangeable lenses so that the photographer could vary the magnification. In all cases, the photographer had to determine in advance the exposure required to obtain a satisfactory image with the available light and the speed or sensitivity of the film emulsion.

The modern history of the camera consists of a series of steps toward total automation. Most consumer cameras of the 21st century are automated, and, in fact, the simpler ones are called *point-and-shoot cameras* (P&S). This degree of automation exceeds anything pioneering photographers could have imagined. We not only have automatic focus and exposure, but higher-end digital cameras come with image stabilization to counteract camera motion during exposure and "face recognition" software so that exposure and focus will be optimized for the areas of most interest in a scene.

In spite of all the automation, knowledge about the way cameras work is still helpful and in some cases essential for high-quality photography. Consider, also, the problem of choosing a camera from the bewildering array of cameras with different features that are available on the market: Why buy a large, expensive camera when a compact, relatively cheap one fits in a shirt pocket and still offers high resolution? The answer depends on the needs of the user, but an informed decision requires some knowledge of the options and trade-offs.

3.3 The Lens

The lens, necessary for forming an image on sensitized film or a sensor in a camera, is characterized by a focal length and an aperture. For simplicity, this chapter discusses only the simple, thin lens.

Definition: The focal length f of a thin lens is the distance behind the lens at which rays from a distant point, i.e., parallel rays, are focused to form an image.

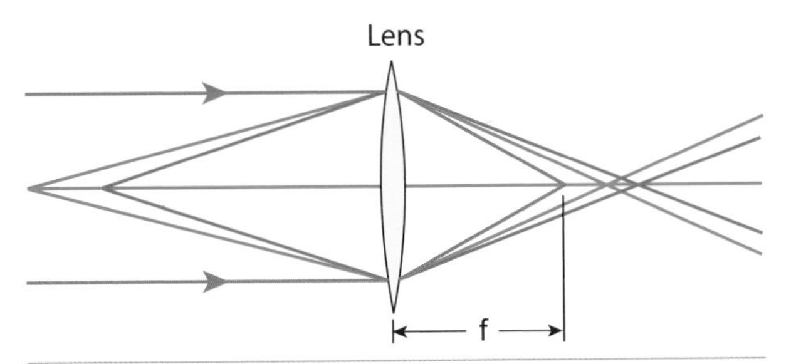

FIGURE 3.2. Parallel light rays from the left focus at the distance *f* behind the lens.

When a cone of light rays from a point on an object not at infinite distance illuminates the lens, the image plane is located a greater distance from the lens than the focal length (see Figure 3.2). The focal length *f* of the lens determines the size of the image and, therefore, cameras with larger film or sensor sizes require lenses with greater focal lengths to capture the same field of view. Chapter 4 explores the concept of the "normal" lens for a camera.

It is interesting to note that the theory of lens design was developed before the advent of photography. In fact, a few well-designed lenses were available before 1839. The earliest lens to be properly designed was the Wollaston landscape lens of 1812, which was still being produced for low-cost cameras in the mid-20th century. The improvement of camera lenses was very slow in the decades after 1840, however, because opticians usually worked by trial and error rather than a systematic application of principles. Lens design is actually a very difficult process that involves many compromises. Light rays must be bent and directed to exactly the correct point on the image, and rays of all colors should be focused at the same point. There are, in fact, seven independent aberrations that need to be corrected to produce a high-quality lens; and the reflections from the glass surfaces also need to be controlled. Fortunately, efficient lens design is now greatly aided by computer programs that can quickly trace light rays through any combination of lenses. Once an acceptable design is found, it can be scaled up or down in size. Of course, the construction of high-quality lenses is still a demanding task, and the required materials such as low-dispersion glass can be very expensive. (Here *dispersion* refers to the way color (actually wavelength) of light affects the focusing ability of glass.)

A photographic lens is said to have a *speed* that determines how much light can be delivered to the image plane in a given time, i.e., the exposure. Lenses are equipped with variable apertures, but each lens has a maximum *aperture*, which is the diameter of the opening that admits light. (The aperture is also called the stop.) Lens speeds are specified by F-numbers, denoted by N or by $f/\#$ where # is the value of N, and the minimum value of N is usually engraved on the lens mount.

Definition: N is the ratio of the focal length, f, to the diameter of the aperture, δ. Thus, we have the formula $N = f/\delta$.

For example, if the focal length is $f = 20$ mm and the maximum aperture is $\delta = 5$ mm, $N = 20\text{ mm}/5\text{ mm} = 4.0$; this is an $f/4.0$ lens, or the lens barrel might be labeled 20 mm 1:4. Notice that the larger the aperture, the smaller the value of N. The F-number is actually a measure of relative aperture, so that a value of N insures the same intensity of light on the sensor regardless of the focal length of the lens. In the special case of close-up photography, the speed or brightness of the lens is diminished because the image plane is farther from the lens than one focal length. An effective N can be defined in terms of the image magnification to take this effect into account (see Chapter 12).

The amount of light admitted to the camera is proportional to the area of the aperture. Recall that, for a circle, area $= \pi(\delta/2)^2$, so when the aperture δ is decreased by a factor of 2, N is doubled and the exposure is decreased by a factor of 4. Note that when the diameter, or aperture, is increased by a factor of $\sqrt{2} = 1.414$, the area doubles. For this reason, the common N's listed on lenses, 2.0, 2.8, 4.0, 5.6, 8, 11, …, correspond to changes by a factor of 1.414, and changes of the exposure by a factor of 2. Each step is called a "stop." Lenses with small N's, i.e., larger apertures, tend to be more difficult to manufacture and are usually more expensive. The larger apertures permit proper exposures to be made with higher shutter speeds (shorter shutter-open times), but there are other trade-offs. Chapter 11 defines the depth of field and shows how it is related to lens aperture.

3.4 Film and Sensors

Nineteenth-century photographers often had to sensitize their own photographic plates and work with hazardous chemicals. In the daguerreotype process, the image was made on silver-plated copper that was coated with silver iodide, and developing the image required warm mercury. By the onset of the American Civil War, the wet-plate collodion process was popular. That method was much less expensive but required the photographer to coat a mixture of *collodion*, gun cotton dissolved in alcohol or ether, with appropriate light-sensitive chemicals on a glass plate, to expose the plate while it was wet, and to develop the plate before it could dry. The rigors of photography during this period limited its appeal to amateurs.

Convenient amateur photography dates from 1895, when Eastman Kodak Company first produced roll film with a paper backing so that it could be loaded in broad daylight. Since that time, numerous film sizes have come and gone, with the dominant surviving ones being 120 and 135 (the standard 35 mm film).[1] These are arbitrary numbers that are not directly related to film size. Cameras that use 120 film make images that are 6 cm in height and range from 4.5 to 9 cm in width. There is also a panoramic camera that uses 120 film and provides an image width of 17 cm. These cameras are all called "medium format." By contrast, 135 film is 35 mm in width, but this includes sprocket holes. The actual image size is usually 24 mm × 36 mm, and this is now called a full-frame (FF) 35 mm image.

In film-based photography, both black-and-white and color, the image is captured in silver halide grains, and the sensitivity of the film is related to the size of the grains regardless of film size. This results from the fact that a few photons activate a grain so that it can be developed to make a silver spot on the negative. The larger grains contribute more to the darkness or density of a negative than the small grains do. Grain is, of course, more noticeable with 35 mm film than with larger films because greater enlargement is required to obtain a desired print size. High sensitivity to light requires large grains, but an image composed of large grains has lower resolution. By resolution, we mean the ability to distinguish closely spaced objects in an image. Other things being equal, one may obtain higher-quality images by using a larger film format, but with the larger format comes a larger, heavier, and usually more expensive camera.

The characteristic of film-based photography is that grains of silver salts sample the intensity of the image projected by the lens, and the developed grains represent the image on film or paper. In comparing film photography with digital photography, it is important to note that film grains are randomly placed. This leads to a sand-like texture in greatly enlarged photographs. When the texture is obvious, the image is said to be "grainy." Color film is much more complicated than black-and-white. The silver grains must be replaced with color dyes, and three layers of these dyes are required for the colors necessary to produce color slides and color negatives. Color film design and the chemistry of image development are discussed in Appendix F.

Digital photography has introduced a range of sensor sizes to replace photographic film. The sensors are microelectronic chips (CMOS or CCD, see Appendix G) that consist of a regular array of detectors that effectively convert photons into electrons so that electrical charge can be accumulated in each picture element. This in turn permits the image to be stored in a computer memory. The area associated with each detector is defined as a *pixel*, which is short for *picture element.*

Definition: A pixel is the smallest element in an image, and can be thought of as a single point in an image. A pixel also represents an area on the camera sensor that contains a light detector.

Pixels are usually square in shape but may be rectangular or even octagonal, as in the super CCDs produced by Fuji. Since chips with large areas are very expensive to make, most consumer digital cameras thus far use very small sensors. Chapter 16 covers sensors in detail. Table 3.1 shows a few typical sensors to illustrate possible pixel sizes and counts. The number of pixels for a given sensor type may vary.

TABLE 3.1.
Typical Sensors for Digital Cameras (2009)

Type	Aspect Ratio	Pixels (millions)	Width (mm)	Height (mm)	Diagonal (mm)
1/2.5"	4:3	8	5.76	4.29	7.18
1/1.7"	4:3	14.7	7.6	5.7	9.6
4/3"	4:3	12.11	18.0	13.5	22.5
APS	3:2	15.1	22.5	15.0	27.04
35 mm	3:2	24.5	36.0	24.0	43.27

The Type column of Table 3.1 contains some arcane nomenclature that is a carryover from TV tube sizes in the 1950s. It is a diameter measurement not simply related to the size of the sensor; it turns out to be about 50% larger than the diagonal of the rectangular sensor. The first two sensors listed are used in compact point-and-shoot cameras; for example, the Canon SD 1100 IS uses the 1/2.5 in. sensor and the Canon G10 the 1/1.7 in. sensor. The relative sensor sizes are shown in Figure 3.3.

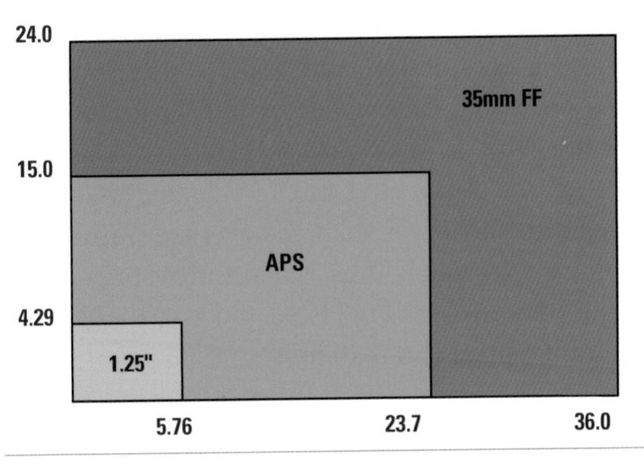

FIGURE 3.3. Relative sizes of some common sensors (mm).

The pixel width (or pixel *pitch*) in the 1/2.5 in. sensor depends on the total number of pixels. For 8.0 megapixels, the pitch is about 1.75 μm. The APS and full 35 mm sensors are found in digital single-lens-reflex (DSLR) cameras. An important difference is that the pixel pitch is much larger in DSLR cameras. It amounts to 4.73 μm and 5.94 μm for the highest-resolution APS and 35 mm sensors listed here, respectively. The difference in pixel areas is even more striking. The ratio of pixel area of the 35 mm sensor to the 1/1.7 in., 14.7 megapixel sensor is approximately 11.4. At the same exposure (light level, lens aperture, and shutter time) a pixel in the 35 mm sensor will receive 11 times as many photons as a pixel in the 1/1.7 in. sensor. It should be noted, however, that the photon detectors are smaller than the pixel pitch. Think of the pixel in a CCD sensor as the square top of a deep box with the detector occupying part of the bottom of the box. In order to increase the efficiency of light detection, a microlens is used in each pixel to focus light onto the detector.

Each sensor type has advantages and disadvantages that go far beyond a simple pixel count. I have already alluded to the difference in signal strength resulting from the number of photons detected. The signal-to-noise ratio is typically lower for a small sensor. In other words, pixels that should be recording the same intensity may show different signals because of fluctuations in the number of photons being detected. The photons arrive randomly just as raindrops do, and the same number does not reach each collector in a given amount of time. This is analogous to catching raindrops in small buckets. The buckets (detectors) may receive more or less than the average number of droplets (photons). Thus, when the average number of photons collected is 36, the actual numbers will range from about 30 to 42 and, in general, the number fluctuation is about equal to the square root of the average number, e.g., 6, or $\sqrt{36}$. The practical consequence of this is that the effective film speed (ISO number) must be kept low for small detectors. (For present purposes the ISO number is just a camera setting that determines how much the signal is amplified. The definition is given in Chapter 16.) This means that larger exposures are required to make up for the small areas of the detectors and to collect comparable numbers of photons in the smaller pixels. As we shall see in Chapter 16, compact cameras produce good low-noise images at low sensitivity, e.g., ISO 100, but above ISO 200 the noise (digital graininess) is often unacceptable. On the other hand, DSLRs that have larger pixels can produce adequate images at ISO 1600 or even higher. The ISO sensitivity setting basically determines the number of photons required to produce a certain signal level.

In practice, neither the graininess of film nor the size of the pixels completely determines resolution. In subsequent chapters, resolution will be defined, and the factors affecting resolution will be explained. In particular, the quality of lenses and the effect of diffraction will be considered. We will see that everything in photography involves compromises. In order to create an image, we face limitations imposed by the laws of physics, the availability of materials, and the present state of technology. The desired properties of the final image must guide our choices of equipment and settings. The appearance of an image depends on the perspective (position of the lens), the field of view, the depth of field, the duration of the exposure, and issues involving quality. Each of these aspects must be examined before we can understand how to obtain equivalent images with different cameras.

CHAPTER 4

Images: What Is Perspective?

When, in early fifteenth-century Florence, Filippo Brunelleschi contrived the first painting in "true perspective," he raised a sensation.

—STEPHEN L. TALBOTT

I like the fact that the word [perspective] literally means "your opinion." Where you place the camera is both an assertion of your camera's viewpoint and your opinion.

—MAHESH VENKITACHALAM

First, let's agree that by *image* we mean a two-dimensional representation of the view defined by the camera lens and the sensor, as shown in Figure 4.1. According to Rudolf Kingslake, a photograph is "a two-dimensional projection of a three-dimensional object, the projection lines all passing through a common center." This brings up the concept of perspective by which we mean the way things appear to the eye based on their positions and distances. Basically, we want to represent a scene in two dimensions so that it gives the same impression of sizes and distances that is experienced in nature.

Some photographs have a natural appearance, while others seem distorted because nearby objects appear too large or distant objects are flattened. These are natural effects having to do with the perspective point of the image and the tilt of the camera. Other distortions, in which straight lines appear to be curved, are different and will be discussed later.

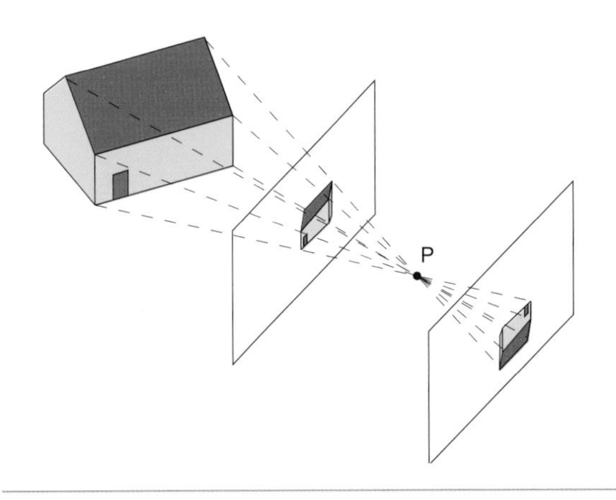

FIGURE 4.1. The perspective point is at *P*, and it may be the eye of the observer or the lens of a camera.

The first thing to recognize is that, for a lens in a particular position, the perspective of the resulting image is independent of the focal length of the lens. For distant objects, however, the size of the image will be approximately proportional to the focal length. This effect, unchanged perspective with a fixed camera position, is shown in Figure 4.2, where the camera lens was changed to obtain focal lengths from 10 mm to 135 mm. Recall that the focal length f of a lens is the distance behind the lens at which parallel rays are focused. The inverse $(1/f)$ is defined as the power of the lens in units of *diopters* when f is measured in meters. This definition will be generalized for thick lenses in Chapter 9. (These comments refer to cameras where the lens plane is parallel to the surface of the detector. Shift and tilt lenses are special cases and require additional analysis.)

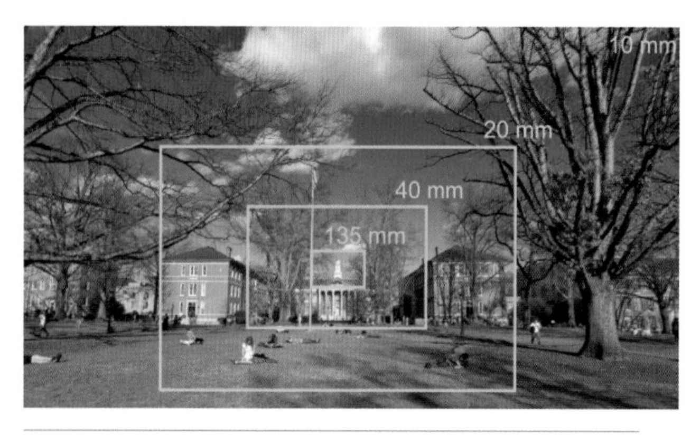

FIGURE 4.2. The effect of changing the focal length while maintaining the perspective. The camera was a Canon XTi (22.2 mm by 14.8 mm sensor).

A normal or natural perspective depends on the angle of view of a photograph relative to that of the original scene. According to Kingslake's analysis, the region of sharp vision of the human eye has an angular width of about 20°, or a half-angle of 10°, and motion is detected over a range of about 50°. The secret to obtaining a normal view is to view a photograph from its perspective point. In that way the eye has the same relationship to objects in the photograph that the lens did to objects in the original scene. The perspective point of a negative or an image sensor is the center of the camera lens or a distance that is approximately the focal length of the lens. (This refers to point P in Figure 4.1) For an enlarged photograph, the perspective distance is the lens distance multiplied by the degree of enlargement.

Suppose, for example, that a 35 mm camera with a 50 mm (about 2 in.) focal-length lens produces a negative (1 in. × 1.5 in.) that is used to make an 8 in. × 10 in. print. The enlargement factor is approximately 8 in./1 in. = 8, and the proper distance from which to view the print is 2 in. × 8 = 16 in. This is a normal viewing situation, but consider a wide-angle lens with a focal length of 17 mm (0.67 in.) that is used with the same camera body. The proper viewing distance for an 8 in. × 10 in. print then becomes 0.67 in. × 8 = 5.36 in. Unfortunately, most 8 in. × 10 in. prints are viewed from approximately the same distance regardless of the point of perspective, and thus the apparent perspective suffers.

So what is a "normal lens"? A normal lens is one that produces photographs that, when viewed normally, will put the eye at the perspective point. A lens produces a circle of illumination and a circle of good definition. Since negatives and sensors are square or rectangular in shape, the diagonal must fit inside the circle of good definition. By definition, a "normal lens" has a focal length approximately equal to the diagonal dimension of the sensor. This choice insures that the camera's angle of view will be similar to the angle of view of the human eye. For a 35 mm camera, the normal lens has a focal length of about 43 mm, and a simple computation shows that the angular (diagonal) field of view has a half-angle of 26.5° [$\theta = \tan^{-1}(0.5)$] or a full angle of 53°. This is somewhat larger than the eye can view at a glance, but it is an acceptable compromise, since prints are usually viewed from beyond the perspective point.

Wide-angle lenses have, of course, shorter focal lengths than normal lenses. A typical wide-field lens has a diagonal angle of view of about 75° or a focal length of 28 mm (FF 35 mm equivalent). Ultra-wide-angle lenses have fields of view from about 90° to 100° and focal lengths of about 17 mm to 20 mm (FF equivalent). These lenses are often blamed for strange-looking, apparently distorted, photographs. The problem is that the proper perspective point for viewing wide-angle prints is inconveniently close to the print. This effect can be alleviated by making larger prints or murals. Figure 4.3 illustrates the change in perspective when the same set of stairs was photographed at close range with a wide-angle lens and from a considerable distance with a telephoto lens.

FIGURE 4.3. A set of stairs photographed with a 10 mm lens (left) and a 190 mm lens (right). The camera was a Canon XTi (22.2 mm × 14.8 mm sensor).

With digital cameras and computer-assisted image processing it is easy to stitch images together to obtain extremely large fields of view. Each image in the resulting panorama usually has a perspective point, and the way to avoid the appearance of an unnatural image is to print or project a very large picture.

There are also narrow field lenses known as "telephoto" lenses. Any focal length longer than the normal focal length will give a telephoto effect, but true telephoto lenses have a compact design in which a negative lens element is inserted to enlarge the image. By definition, the length of a true telephoto lens from the front element to the film plane is less that the focal length.

Of course, most cameras for amateurs are sold with zoom lenses. These lenses smoothly change focal length with a twist, a push/pull motion, or the pushing of a button. This will not change the perspective of the image, but it will certainly change the perspective point from which photographs should be viewed. Fixed focal length lenses are known as prime lenses.

Our final point about perspective involves camera tilt. It is common knowledge that when a camera is tilted up, the image will be distorted so that vertical lines tend to converge, buildings appear to lean backward, etc. This effect is often considered undesirable, and methods are sought to avoid or to correct for it. Holding the camera perfectly level will remove the distortion, but may clip the tops of objects and may miss points of interest. Professional view cameras avoid the tilt problem by raising the lens while holding the lens plane parallel to the film plane, and there are special-purpose tilt/shift lenses available for interchangeable lens cameras, e.g., the Canon TS series.

Corrections for tilt in existing negatives can sometimes be made by tilting the paper holder in a photographic enlarger. Enlargers are less common in the digital age, but much more powerful tools exist to take their place. Image-processing software such as Photoshop can transform (distort) an image to correct for the convergence of lines. Computer programs can also correct for lens aberrations and distortions and can even convert images taken with fisheye lenses to rectilinear form.

Strangely enough, extreme camera tilt is often considered pleasing. Also, sideways tilt that makes horizontal lines converge at the horizon is not considered a defect. This effect is so common and accepted that artists often use geometric projections to build this kind of perspective into their images. In this and all the other cases mentioned, a natural view is obtained by using one eye at the perspective point. The use of two eyes (binocular vision) gives an unnatural appearance for small images.

We don't notice this problem with natural scenes, and we can avoid it by making large images.

In closing this discussion I note that our perspective point is determined when we select a seat in a movie theater, but that position is seldom the perspective point of the movie photographer.

Further Reading

R. Kingslake. *Lenses in Photography: The Practical Guide to Optics for Photographers.* Rochester, NY: Case-Hoyt Corp., 1951. (See Chapter 1).

R. L. Solso. *Cognition and the Visual Arts.* Cambridge, MA: MIT Press, 1994. (See Chapter 8.)

CHAPTER 5

Why Does a Camera Need a Lens?

Look and think before opening the shutter.
The heart and mind are the true lens of the camera.

—YOUSUF KARSH

I think the best way to discover the need for lenses is to analyze a famous lensless camera, the pinhole camera. This is nothing more than a small version of the camera obscura that played such a large role in the history of art and photography. All we need is a light-tight box with a pinhole in the center of one wall and a photographic plate attached to the opposite wall. As shown in Figure 5.1, the formation of an image can be explained with nothing more than ray optics. That is, we assume that light beams move in a straight line, and we perform ray tracing. A ray of light from any point on an object passes through the pinhole and strikes the plate on the opposite wall.

It is obvious from the figure that a point on the object will become a small circle in the image because the pinhole aperture must have some diameter to admit the light. If we make the pinhole smaller, the light reaching the image will be diminished, and the exposure will take longer. If we enlarge the pinhole, the resolution will suffer. At first

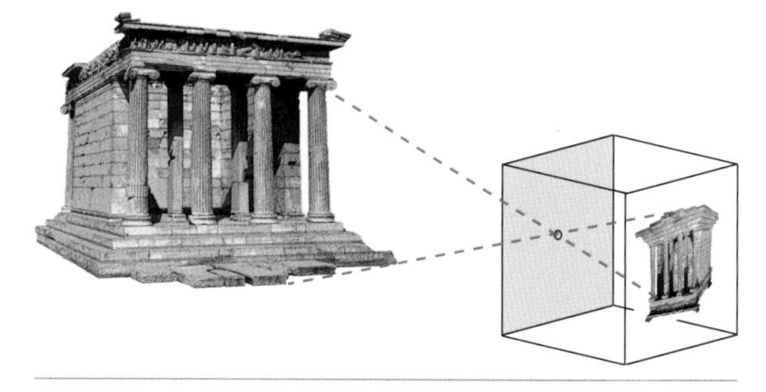

FIGURE 5.1. Illustration of a pinhole camera aimed at the Temple of Wingless Victory (Athens).

it appears that there is no optimum size, and we just have a trade-off between resolution and the time required to expose the plate. But that turns out not to be the case; if we try smaller and smaller pinholes, at some point we find that the resolution actually decreases. The reason for this development is the ubiquitous diffraction phenomenon.

In the limit of very small apertures, the image of a distant point, say a star, becomes a circular disk with faint rings around it. This bright central region, called the Airy *disk*, is shown on the right side of Figure 5.2. The diameter of the disk measured to the first dark ring is equal to $2.44\lambda N$, where λ is the wavelength of the light and N is the ratio of focal length to aperture, f/δ, defined in Chapter 4 as the F-number. The spot size for a distant point without diffraction is just equal to the aperture diameter δ. Therefore, the spot size will increase for both large and small pinholes.

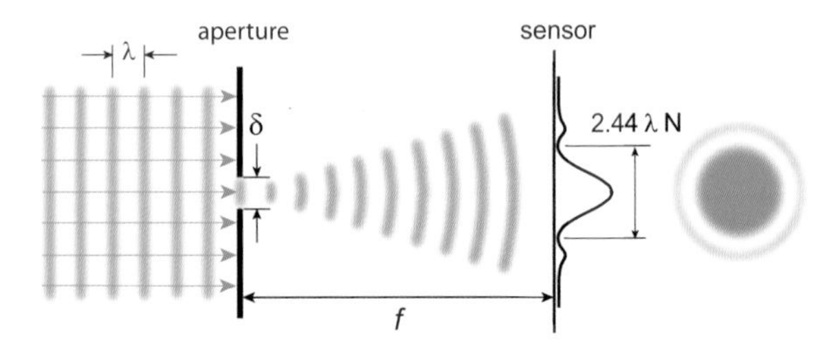

FIGURE 5.2. The effect of a small aperture on light incident from the left. The vertical scale is greatly exaggerated to reveal the intensity distribution in the diffraction spot.

The optimum pinhole size occurs when the contributions of these two effects are approximately equal. Therefore,

$$\delta = 2.44\lambda f/\delta, \tag{5.1}$$

and we conclude that the optimum aperture size is equal to

$$\delta_{opt} = \sqrt{2.44\,\lambda f} \tag{5.2}$$

Since $\lambda \approx 500$ nm in the middle of the visible spectrum, we find on average that

$$\delta_{opt} = 0.035\sqrt{f} \tag{5.3}$$

for δ_{opt} and f measured in millimeters. When $f = 100$ mm, the optimum pinhole has diameter $\delta_{opt} = 0.35$ mm. This is approximately the size of a no. 11 needle. A formula attributed to Lord Rayleigh improves this calculation by the substitution of 3.61 for 2.44 in Equation (5.1).

This derivation is only approximate, and it fails badly for nearby objects. It is clear that light rays from a bright point near the pinhole will diverge through the pinhole to make a spot on the film that is larger than the pinhole. This results in a loss of resolution for nearby objects, and the realization that the optimum size for the pinhole depends on the distance from the pinhole to the object being photographed.

Photography with a pinhole camera requires an exposure time that depends on the F-number (N) of the aperture. From the definition we find that $N =$ 100 mm/0.35 mm or $f/286$ for this camera with its optimum pinhole size. How does this compare with a typical lens-based camera? A typical aperture setting for outdoor photography is $f/8$, and we have found that every time N doubles, the intensity is cut by a factor of 4. Thus $f/8$ must be doubled five times to equal approximately $f/286$, and this indicates about $4^5 = 1024$ times as much light intensity on the image plane as with the pinhole camera. The recommended exposure time for typical (ISO 100) film in hazy sunlight is 1/125 s at $f/8$. The corresponding exposure for a pinhole camera with the optimum pinhole would be about 8 s.

Thus, a pinhole camera gives at best poorly resolved images and requires very long exposure times as well. It does produce interesting effects, since any focal length is possible and the images show no aberrations or distortion. The long exposure times can also produce unusual effects. For example, during long exposures objects can move quickly through the field of view and leave no trace in the image. In fact, there is a lot of interest in pinhole cameras by hobbyists, but pinhole cameras are toys compared to cameras with lenses.

One other point is worth noting. The ray-tracing approach described above assumes that the rays are unaffected by the pinhole other than the spreading effect of diffraction. This works if the same medium, such as air, fills the camera and the outside space. Suppose instead that the space between the pinhole and the film is filled with water. A ray passing from air to water is subject to refraction, and the path is bent toward the optical axis (a line through the pinhole and normal to the film plane). The unexpected result is a fisheye view of the world. This effect is discussed and illustrated in Chapter 10. An air-filled pinhole camera with a glass window underwater would be the reverse situation and should give a telephoto effect. As far as we know, this arrangement has not been demonstrated.

In summary, here is what lenses do. They permit sharp, high-resolution photos to be made with large apertures. The large apertures mean that adequate exposures can be obtained with high shutter speeds (short exposure times). In addition, the effects of diffraction can be reduced and the region of sharp focus (depth of field) can be controlled. Thus, we see that lenses have transformed a curiosity into an image-making miracle.

Further Reading

J. Grepstad. "Pinhole Photography." Available at http://photo.net/learn/pinhole/pinhole, 2009.

CHAPTER 6

Elementary Optics: How Do Lenses Work?

Now, let's have some fun. Let's "fool the light," so that all the paths take exactly the same amount of time.

—RICHARD P. FEYNMAN

I will begin with a few basic rules of optics and demonstrate how they work in simple lenses. Of course, science is not just a list of independent rules, and it should be recognized that these rules are derived from more general laws of physics. The rules are not only consistent with Maxwell's electromagnetic theory, but they can be derived from the quantum theory of photons and electrons. Showing how all that works is the fun part for those who enjoy physics and mathematics. Appendix B presents an introduction to these ideas.

The most important fact is that light travels at the speed $c = 2.997 \times 10^8$ m/s in a vacuum such as outer space, but the speed appears to be reduced in transparent matter, such as water or glass.[1] Each material has a characteristic index of refraction n that is greater than one, and the apparent speed of light in the material is equal to c/n. The refractive index depends on the frequency of the light and conditions, such as temperature and pressure, that affect the properties of the material. Refractive indices have been measured for all the materials of interest in optics and can be found in tables. For water at 25° C, the refractive index is 1.33, and, therefore, the apparent speed of light in water is reduced to $c/n = 2.253 \times 10^8$ m/s.

Here are the rules we need for the optics of light beams (rays):

1. In a material with a uniform index of refraction, light goes from place to place in a straight line.
2. When light is reflected from a surface, the angle of incidence, denoted by θ_i, is equal to the angle of reflection, denoted by θ_r. See Figure 6.1.
3. When light passes from a medium with refractive index n_1 into a medium with refractive index n_2, the angles of incidence, θ_1, and refraction, θ_2, are related by a formula known as Snell's law:

$$n_1 \sin\theta_1 = n_2 \sin\theta_2 \tag{6.1}$$

where sin refers to the sine function of trigonometry. The geometry of refraction is shown in Figure 6.2.

4. Diffraction effects must be considered when light interacts with an aperture with a diameter that approaches the wavelength of the light. (This is not really a rule, but rather a warning that ray optics do not tell the whole story.)

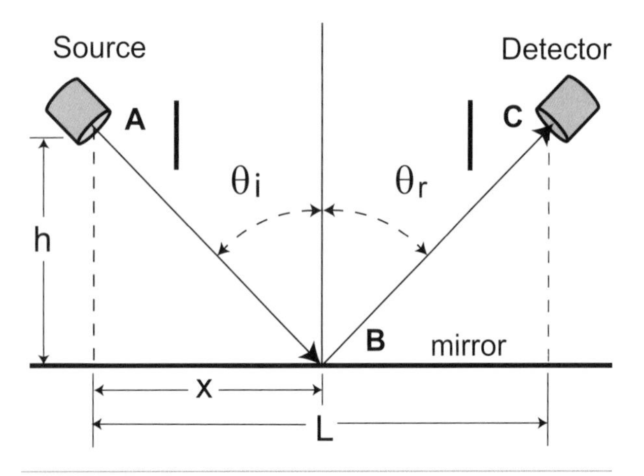

FIGURE 6.1. Illustration of Rule 2, equal angles of incidence and reflection.

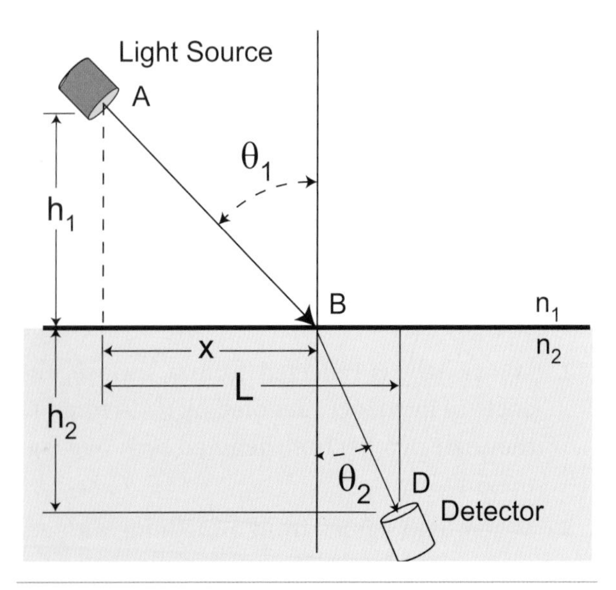

FIGURE 6.2. Illustration of Rule 3, Snell's law of refraction.

The rule that light travels in a straight line is essentially proved correct by the pinhole camera when the effects of diffraction are neglected. So we will assume that it is correct and will use it in considering the other rules. The second rule is known as Hero's Law of Reflection because he was the first person to publish it. This is Hero (or Heron) of Alexandria, who was a mathematician in the 1st century CE. He is known for inventing the steam engine, but he was also interested in optics and he studied mirrors, probably polished metal or the surface of water. Hero's Law can be derived from Fermat's *principle of least time*, light of a given frequency moves between two points along the path that takes the least time. Pierre de Fermat was one of the best mathematicians in the world in the mid 16th century. He is best-known now for his "last theorem," which was proved only in the 1990s.

The principle of least time is applicable to many problems in optics and deserves our attention. First, note that in a uniform medium, where the index of refraction is constant, the path that requires the least time is also the shortest path. If the path traverses regions where the refractive indices are different, then the path requiring least time must be chosen. This also leads to the idea that the optical path length is equal to the physical path length multiplied by the appropriate index of refraction.

In the reflection problem shown in Figure 6.1, the path from A to C, i.e., path ABC has a minimum length when $x = L/2$. From the Pythagorean Theorem we find that the path length can be expressed as

$$P(ABC) = \sqrt{h^2 + x^2} + \sqrt{h^2 + (L - x)^2} \tag{6.2}$$

It is easy to verify that the minimum value of $P(ABC)$ occurs when light reflects from the mirror at the midpoint between A and C, and that the value increases when any value of x other than $L/2$ is used. So the trial-and-error method works here, but the straightforward way to obtain the minimum path is to use the calculus.[2]

Rule 3, Snell's law, is the important one for lens design. Willebrord Snellius was a professor of mathematics at the University of Leiden when he discovered the law of refraction in 1621. For some reason he never published the law, and it only became known in 1703 when Christiaan Huygens revealed the equation in his *Dioptrica*. Snell's equation can also be derived from Fermat's principle of least time. In Figure 6.2, the light beam passes from medium 1 (air) to medium 2 (glass). We will assume that $n_1 < n_2$, which is appropriate for air ($n_1 = 1$) and crown glass ($n_2 \simeq 1.55$). The figure shows that the path length in air has been increased relative to the path length in glass, and this is necessary to minimize the time, since the speed in air (c/n_1) is greater than the speed in crown glass (c/n_2). This is analogous to the situation of a lifeguard who must rescue a swimmer who is down the beach and out to sea. Since the guard can run faster than he can swim, he should not head directly for the swimmer but should run at an angle so that the distance covered on the beach is greater than the distance in the water and the total time can be minimized.

The amount of time required for a light ray to go distance L in a medium having the refractive index n is $t = L/(c/n) = nL/c$, and we call nL the optical path length (as opposed to the physical path length L). Therefore, the time required for the light ray in Figure 6.2 to cover path ABD is given by

$$t = \left[n_1 \sqrt{h_1^2 + x^2} + n_2 \sqrt{h_2^2 + (L - x)^2} \right] \Big/ c. \quad (6.3)$$

The minimum value of t can be obtained by calculus or by a trial-and-error variation of x. It turns out that the minimum value of t is obtained when

$$\frac{n_1 x}{\sqrt{h_1^2 + x^2}} = \frac{n_2 (L - x)}{\sqrt{h_2^2 + (L - x)^2}}. \quad (6.4)$$

Equation (6.4) is identical to Snell's Law as described in Equation (6.1), since

$$\sin\theta_1 = x \Big/ \sqrt{h_1^2 + x^2}$$

and

$$\sin\theta_2 = (L - x) \Big/ \sqrt{h_2^2 + (L - x)^2}.$$

Now consider the simple lens shown in Figure 6.3. Parallel light rays emitted from a distant source on the left are focused on the image plane at a common point. According to Fermat's principle of least time, this happens because all of the rays go from the source to the focal point in the minimum time. This is possible because the apparent speed of light is lower in the lens, i.e., c/n is less than c. The ray through the center of the lens has the shortest path in air but is delayed by passage through the thickest part of the lens, and so on, and the lens is designed so that the optical path lengths and times for all of the rays shown are identical. From the starting points on the left-hand side, other paths to the desired focal point would require more time.

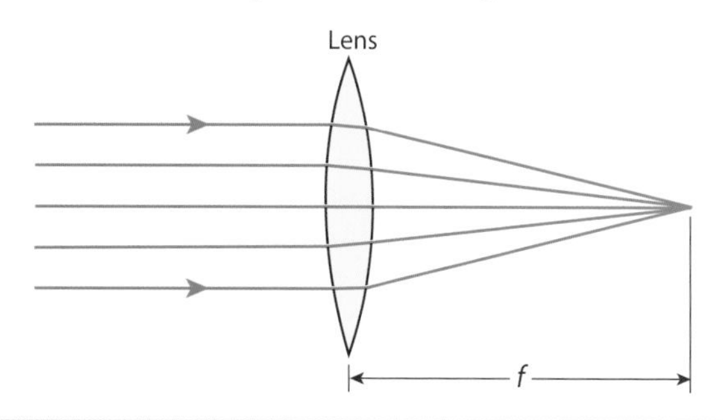

FIGURE 6.3. Light rays and a simple lens. Since the speed of light is lower in a glass lens than in air, all the paths take the same time to reach the focal point.

The principle of minimum time is a unifying principle that justifies, and can replace, the simple rules for reflection and refraction. But, does this principle always apply? Unfortunately, the answer is no. There are cases where light rays appear to seek the path of maximum time. So there is more to study and understand. The important idea behind all of optics is that a light ray is composed of photons but acts like a wave. In the classical picture, described by Maxwell's equations, the wave represents the amplitude of the oscillating electric field associated with the light ray, and the square of the amplitude is proportional to the intensity of the light. In the quantum picture, the light ray is described by a wave function, and the square of the absolute magnitude of this function gives the probability of finding a photon.[3] This is all consistent because the intensity is determined by the number of photons.

So, fine, everyone agrees that we have waves, but how does this idea generate the rules of optics? The answer is the *superposition principle*; namely, the amplitude of a light wave is the sum of the amplitudes of all light waves at the same time and position. The waves associated with all the relevant light rays must be added to obtain the resultant wave before the amplitude is squared. It is only the set of rays that take the least time (or possibly the maximum time) that are in phase and contribute significantly to the intensity. This simple idea, which is explored in Appendix B, is the topic of Richard Feynman's beautiful little book, *QED, The Strange Theory of Light and Matter.*

CHAPTER 7

The Simple Thin Lens and What it Does

A thin lens is a design tool used to simulate a real lens during the preliminary stages of optical system design and analysis.

—BRUCE WALKER

7.1 Introduction

The earliest lenses were made of polished crystals, and it was not until the middle ages that glass lenses were produced. By the 13th century, glass lenses were good enough to be used in spectacles for the correction of presbyopia. More complicated optical instruments such as telescopes require lenses of sufficient uniformity that they can be used in pairs to obtain magnification. The master spectacle maker Hans Lipperhey of Middelburg improved lens grinding techniques and in 1608 was able to present a telescope to Prince Maurits in The Hague.[1] Others claimed priority in inventing the telescope, but the first record we have of an actual instrument appears in Lipperhey's patent application. The time was right, ideas were in the air, and someone was going to produce good lenses and construct a telescope in the early 17th century. Unfortunately, Lipperhey didn't get his patent, but he was well paid for his services, and he now has a lunar crater and a minor planet named for him.

Everything changed in 1610, when Galileo made a telescope for himself and turned it on the heavens. He quickly discovered the moons of Jupiter and published his results. That same year Kepler got his hands on a telescope made by Galileo and worked out a new theory of optics based on two ideas:

1. Light rays travel in all directions from every point in an object (Maurolico, 16th century).
2. A cone of light rays enters the pupil of the eye and is focused on the retina (Kepler, 1604).[2]

The revolutionary developments of 1610 opened a new world for scientific observations of objects both near (by the microscope) and far (by the telescope).

The goal of this chapter is to show how points in subject/object space (in front of a lens) are related to points in the image space (behind a lens) for ideal lenses. With modern equipment it is relatively easy to grind and polish the surface of a circular glass plate into a spherical shape, that is to say, having the shape of a portion of the surface of a sphere. In a plane that contains the optical axis (axis of symmetry), a spherical surface appears as a segment of a circle. To this day, most camera lens elements are spherical in shape. We show in Chapter 8 that aspheric surfaces are sometimes desirable, but they are much more difficult to fabricate. The next step is to show how a lens can be shaped so that it will have a certain focal length. This is, in fact, the starting point for understanding the design of all lenses, even compound, multi-element lenses.

7.2 Optical Surfaces

No matter how complicated a lens is, it still consists of optical surfaces separating regions that have different indices of refraction; and, in principle, all we need is Snell's law of refraction to be able to trace the path of a ray of light through a lens. This idea is illustrated in Figure 7.1, where a ray of light (red) is incident on a spherical surface at a distance h from the optical axis. The surface is characterized by the radius of curvature R; since the center of curvature is on the right side, the surface is convex and we assign the radius a positive value. This sign convention must be applied consistently; if the center were on the left side, the surface would be concave to the incident ray, and the radius would be negative. The effect of the surface is to change the direction of a ray if the refractive indices on the two sides (here n_1 and n_2) are different. The calculation of the change in angle from θ_1 to θ_2 relative to the surface normal just requires the application of Snell's law.

Snell's law (Chapter 6) tells us that $n_1 \sin \theta_1 = n_2 \sin \theta_2$, and it is very easy to compute the angle of refraction for a single ray passing through a surface. The calculation must, of course, be repeated for each surface of the simple lens and many surfaces in a compound lens. Furthermore, in order

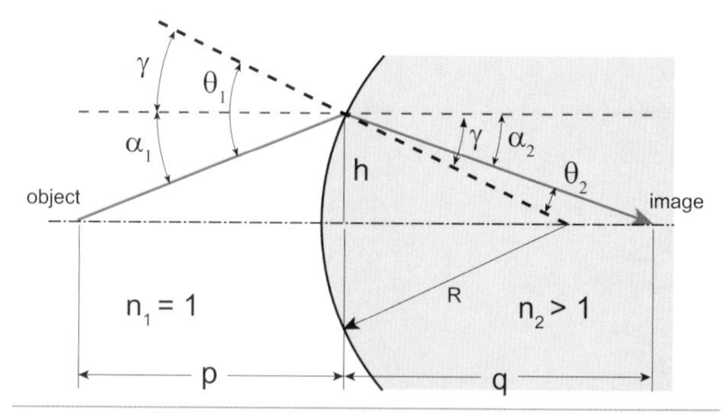

FIGURE 7.1. A spherical surface with radius R. The dotted line from the center of curvature is normal to the surface.

to characterize a lens, a representative set (cone) of rays from points on the object side must be used to sample the entrance surface of the lens. This is a tedious task, and nowadays it is easily performed by a computer. Such a set of calculations can predict the performance of a particular lens design but does not give much guidance about how to design a lens. For that we need simple equations and rules of thumb to get started. Here, as in many areas of science, simplifying assumptions can lead to very useful approximate equations. For example, suppose that the rays remain close to the optical axis and that all of the angles are small. This means small enough that the sin θ and tan θ functions can be replaced with θ itself, where the angles are measured in radians (2π radians $= 360°$). This is called the paraxial approximation with which Snell's law becomes $n_1\theta_1 = n_2\theta_2$.

With the paraxial approximation it is easy to derive equations for a simple thin lens. The optical power of the surface shown in Figure 7.1 is defined as $P_s = (n_2 - n_1)/R$, and our first goal is to relate this power to the object and image distances p and q, respectively. Consider the situation shown in Figure 7.2. Notice that the radius for the first (entrance) surface R_1 is greater than zero as in Figure 7.1 because the center of curvature is on the extreme right-hand side, and the radius of the second (exit) surface R_2 is less than zero because its center of curvature is far on the left-hand side. The first step is applying Snell's law to derive an equation for the focusing power of surface 1. We inspect Figure 7.1 and find that $\theta_1 = \alpha_1 + \gamma$ and $\theta_2 = \gamma - \alpha_2$. Then the paraxial approximation permits us to use the following expressions for the angles: $\alpha_1 = h/p_1$, $\alpha_2 = h/q_1$, and $\gamma = h/R_1$. When these quantities are substituted into Snell's law, we obtain the following equation for the power of surface 1:

$$P_s = \frac{n_2 - n_1}{R_1} = \frac{n_1}{p_1} + \frac{n_2}{q_1}. \quad (7.1)$$

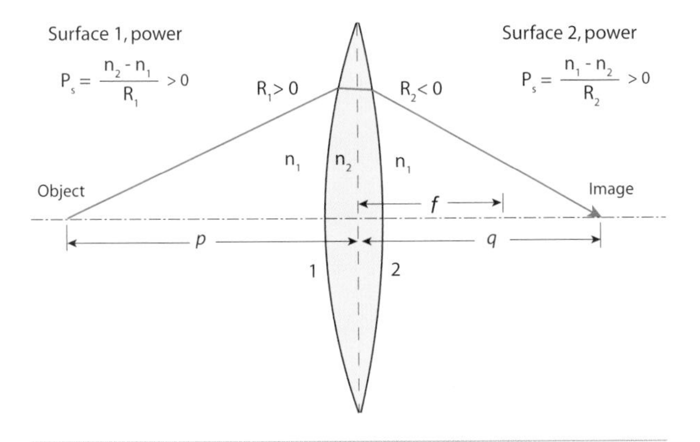

FIGURE 7.2. A simple convex–convex spherical lens.

Equations that describe the thin lens require that the powers of the two surfaces be combined. These derivations, which can be found in Appendix C, yield

$$\left(\frac{n_2}{n_1} - 1\right)\left(\frac{1}{R_1} - \frac{1}{R_2}\right) = \frac{1}{p} + \frac{1}{q}. \quad (7.2)$$

If the space before the lens is occupied by air, so that n_1 is close to 1, the focusing power depends only on the refractive index of the lens material and the radii of curvature. If the entrance surface of the lens is in Ω with water, which has a refractive index of approximately 1.3, then the effective power will be reduced. It can, of course, be compensated by decreasing the radius of curvature. Actually, we assume that the refractive indices in the object and image spaces are equal, and, in general, they are set equal to 1.

Now suppose that the object is moved very far to the left so that $1/p$ approaches zero. In this limit, q becomes equal to the focal length f, and Equation 7.2 becomes the well-known *lensmaker's equation*,

$$\frac{1}{f} = \left(\frac{n_2}{n_1} - 1\right)\left(\frac{1}{R_1} - \frac{1}{R_2}\right). \quad (7.3)$$

In Equation (7.3) lensmakers have a useful recipe for relating the properties of glass and the shapes of lens surfaces to the focal lengths they require. Another useful equation, (7.4), is obtained by noting that the left-hand side of (7.2) is equal to the right-hand side of (7.3). Thus,

$$\frac{1}{f} + \frac{1}{p} = \frac{1}{q}. \quad (7.4)$$

This is called the *conjugate equation* since it relates the conjugate distances p and q to the focal length f. Equation 7.3 is wonderful for lens designers, but is usually not of much interest to photographers. On the other hand, the conjugate equation is very useful for understanding the way lenses work. Right away we see that a simple lens can be focused by adjusting the lens-to-sensor distance q to accommodate the object-to-lens distance p. Furthermore, the magnification (the size of the image of an object relative to the actual size of the object) is equal to the ratio q/p and is, therefore, determined by the focal length. When the object distance p is much greater than the focal length f, the magnification is approximately equal to f/p. Note that when $p = q$, the magnification m is equal to 1 and that both p and q are equal to $2f$. These topics are explored in greater detail in Chapter 12.

Caveat. In the paraxial limit the thin lens is ideal in the sense that a cone or rays from a point in an object will be focused into a circular cone of rays that converges to a point (or to a diffraction limited disk). As the diameter of the lens (or the limiting aperture) increases so that the paraxial no longer applies, focused bundles of rays deviate from circular cones and take on a variety of patterns. The patterns are characteristic of spherical lenses and are associated with well-known aberrations. Chapter 8 covers aberrations and ways to minimize their effects.

7.3 Compound Lenses

The focusing power of the lens, taking into account both surfaces, is characterized by the quantity $1/f$. Now, suppose we have two thin lenses in contact as in Figure 7.3. What is the power of the combination? This is an easy question that can be answered by combining the conjugate equations for the two lenses. First we introduce the labels 1 and 2 to obtain

$$\frac{1}{p_1} + \frac{1}{q_1} = \frac{1}{f_1}$$

and

$$\frac{1}{p_2} + \frac{1}{q_2} = \frac{1}{f_2} . \tag{7.4}$$

The image for the first lens falls to the right of the second lens and becomes its subject. Therefore, we can substitute $-q_1$ for p_2 in the second equation where the minus sign indicates that the subject is behind (to the right of) the lens. The equations can then be added to obtain

$$\frac{1}{p_1} + \frac{1}{q_2} = \frac{1}{f_1} + \frac{1}{f_2} . \tag{7.5}$$

The subscripts on the left-hand side can be dropped because the combined set of thin lenses only has one subject (object) distance and one image distance. The conclusion is that the focusing power of a set of lenses can be computed by adding their separate powers. From Chapter 6 we know that the quantity $1/f$ is defined as the power in diopters when f is measured in meters, and the rule is to add diopters when lenses are used in combination.

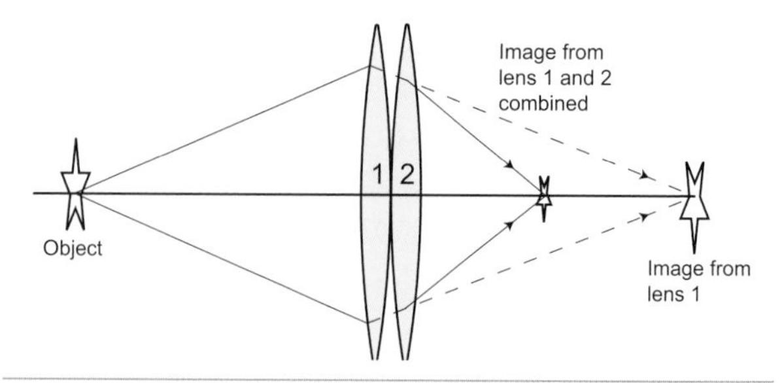

FIGURE 7.3. A pair of thin lenses with exaggerated thicknesses.

A more involved derivation shows that when the two lenses are separated by a distance d, the equation becomes

$$\frac{1}{p_1} + \frac{1}{q_2} = \frac{1}{f_1} + \frac{1}{f_2} - \frac{d}{f_1 f_2} . \tag{7.6}$$

7.4 Conclusion

The simple thin lens provides a useful model for understanding all lenses. The conjugate equation is the starting point for estimating the behavior of lenses and even applies to compound lenses under conditions that are discussed in Chapter 9.

Further Reading

V. Ronchi. Optics: *The Science of Vision.* New York: Dover, 1991.

G. R. Fowles. *Introduction to Modern Optics, Second Edition.* New York: Dover, 1989. (This is a good modern optics text.)

M. W. Davidson. "Bi-Convex lenses." Molecular Expressions. Available at http://micro.magnet.fsu.edu/primer/java/lens/bi-convex.html. 2009.

CHAPTER 8

How to Make Lenses that are Good Enough for Photography

Nevertheless, it remains true that
the lens is generally the most expensive and
least understood part of any camera.

—RUDOLPH KINGSLAKE

Every defect and aberration that a lens has
will be revealed in an astrophoto. These include spherical
aberrations, chromatic aberrations, coma, astigmatism,
curvature of field, and vignetting.

—JERRY LODRIGUSS

8.1 Introduction

After photography was invented in 1839, there was a pressing need for lenses that could produce sharp, undistorted images with wide fields of view. The need was much less obvious when objects were viewed through the center of telescopic lenses and artists squinted at faint images on ground glass in a camera obscura, but photography provides permanent images that can be examined critically in bright light. The new requirements seem straightforward. We want all the light from an object point to be focused in the correct position on the film/sensor, regardless of color. It turns out to be possible to make an acceptable image but not a perfect image. Here is the problem: We are limited by the laws of physics, the properties of available optical glass, and the technology for shaping glass surfaces. In this chapter we consider why multi-element lenses are necessary in order to minimize aberrations while providing large apertures. We will also find that multi-element lenses give unacceptable light transmission, ghosts, and flare in the absence of antireflective multi-coating.

All the possible types of spherical lenses are shown in Figure 8.1.

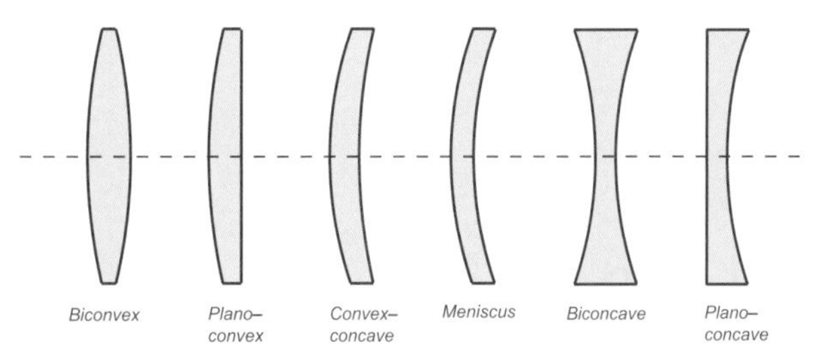

FIGURE 8.1. Simple spherical lenses.

As discussed in Chapter 6, those lenses that are thicker in the center (positive lenses) can be used to focus light, while the concave lenses (negative lenses) cause light rays to diverge. Suppose we select a single biconvex lens for our camera. The resulting image will be something like the one shown in Figure 8.2. It may be possible to get a sharp image in the center, but the resolution will deteriorate away from the center. The main problem is that the image projected by the biconvex lens is not flat, but there are other aberrations as well.

FIGURE 8.2. Photograph illustrating the effect of image curvature (simulated).

In 1812, the English scientist W. H. Wollaston discovered that a meniscus-shaped lens, actually convex–concave, with its concave side to the front, can produce a much flatter image and, therefore, a much sharper photograph overall. Unfortunately, Wollaston's simple lens is not suitable for photography because it exhibits extreme chromatic aberration, i.e., it focuses blue light at a different place on the optic axis than it does yellow light.

The failure of biconvex and convex–concave lenses in photography is just an example of the host of problems that must be faced when attempting to improve the photographic lens. Major problems arise because it is relatively easy to produce lenses with spherical surfaces by grinding and polishing, but spherical lenses are not optimal. In fact, the major monochromatic aberrations (spherical aberration, coma, astigmatism, and distortion) result from the inability of spherical lenses to produce perfect images. If one has to use spherical lenses, the correction of aberrations requires a compound lens with multiple elements.

Science advances by erratic steps. At times inventions are followed by the development of a theory that leads to substantial improvements. For example the steam engine was followed by the theory of thermodynamics. Thermodynamics then showed how to make a more efficient steam engine. In electronics, the order was reversed. Maxwell's unifying equations came first, after which Hertz produced radio waves. In optics, we find a mixed bag. The great theoreticians, Gauss, Petzval, von Seidel, and others, developed the theory of lenses, but many lensmakers muddled along hoping to find a great lens design by trial and error (see [Kingslake 89, Chapter 1]). Consequently, the development of good camera lenses was a very slow process. The problem is that aberrations cannot be corrected without using a number of spherical lens elements. This means that a lensmaker must adjust the radius of curvature of the surfaces of each lens element, the thickness of the elements, their separations, the position of the aperture (stop), and even the type of glass in each lens element. Without some guiding principles, the lensmaker is lost.

Everything in photography involves compromise. We can never make a perfect lens, but we can make lenses that are optimized for some purpose, taking into account size and cost restrictions. Therefore, lens imperfections are still with us, albeit minimized by advances in computation and materials science, and it is instructive to study the criteria used to evaluate modern lenses. Basically, we expect the following:

- sharp image over the entire frame (low aberrations with high resolution and contrast),
- undistorted image (limited barrel and pin cushion distortion),
- no colored fringes around objects (little chromatic aberration),
- uniform brightness over the image (little vignetting),
- little scattered light and good light transmission (low light loss).

Next, we explain the terms in parentheses and describe remedies. The first three items are related to lens aberrations, though they are measured in different ways and usually discussed separately. Correction requires the combination of a number of spherical elements, some of which have a different index of refraction. The modern lens designer knows lens theory and has access to catalogs of successful designs from the past. A new lens design can be based on simple calculations or possibly on a previous successful design, but the task of optimizing all the elements is done by computer. A modern lens design program such as OSLO (optics software for layout and optimization) requires the user to enter information about the lens elements with a specification of glass types and the parameters that are to be held constant. Various parameters such as the curvature of lens elements are specified as variables. The program then determines all the optimum parameters (for minimizing aberrations) by tracing a large number of light rays through the lens system. With such programs, lens performance can be measured without having to construct a real lens.

In lens design some things are impossible to do with a given set of materials. A very-wide-angle lens with a large aperture will always have some image distortion, and the resolution will be better in the center (on-axis) than at the edges. Similarly, zoom lenses introduce a new set of compromises. Usually, there will be some barrel distortion (outward ballooning of the edges) at the shortest focal length and some pincushion distortion (pinching-in of the edges) at the longest focal length. The wider the zoom range the more difficult it will be to maintain high resolution and low distortion for all focal lengths. One should keep in mind that a lens designer must decide what compromises to make given the limitations of weight, size, and cost. Users can now find lens reviews online and can decide what they are willing to pay for and which trade-offs they are willing to make.

8.2 Aberrations

The term aberration appears so frequently in the discussion of lenses that it is worthwhile to be careful with its definition (see [Klein 70, Chapter 4] and [van Walree 09]). It is necessary to recall that the refraction of light rays at each optical surface is properly described by Snell's law. For those rays that are close to the optical axis and make small angles with it, the Snell's law calculations can be greatly simplified by replacing the trigonometric function, $\sin\theta$, with θ itself. As we saw in Chapter 7, this substitution amounts to what is called *paraxial* or *first-order theory* and the concept of the perfect lens where $\sin\theta = \theta$ is satisfied. With practical lenses, however, this approximation is not justified and something better is required. In 1857, Ludwig von Seidel turned to the next best approximation of $\sin\theta$ when θ is still small, namely $\sin\theta - \theta - \theta^3/6$, and he worked out so-called third-order optics in order to see what changes result from the extra terms. His calculation, still an approximation, revealed five independent deviations from the perfect lens behavior for monochromatic light. These deviations are known as *Seidel aberrations* or more simply *aberrations*. With larger angles more extreme effects appear that are not really independent, but third-order optics provides a vocabulary and a good starting point for lens design. (For a complete listing of the aberrations with diagrams please visit Wikipedia[1] and the Melles-Griot websites.)

It is interesting to consider how spherical aberrations can be corrected by using only spherical lens elements. Suppose that a hypothetical aberration-free lens focuses incident light rays that are parallel to the optical axis at a distance f behind the lens. When longitudinal spherical aberration (LSA) is present, parallel rays off the optical axis are focused at distances less than f and the marginal rays serve to define the LSA. If the light is monochromatic, and the spherical aberration is small (described by third-order theory), it can be shown that the resulting spot size for a light ray is given by kf/N^3, where k is the aberration coefficient that must be determined (or perhaps looked up in a catalog), f is the focal length of the lens including the sign (+/−), and N represents the f-stop. Thus, in order to improve the optical performance, we need to add elements so that the total aberration for the system can be minimized. For example, a positive lens $f_1 > 0$ can be used in combination with a negative lens $f_2 < 0$ with the f_1/f_2 ratio adjusted to cancel the aberration. If other aberrations such as astigmatism, coma, field curvature, and distortion are present, still more lens elements are required.

If all these aberrations result from the use of spherical lenses, why not just use aspheric lenses? That is a good idea because aspheric lenses with appropriate shapes can eliminate spherical aberrations and cut down on the number of elements required in compound lenses. Unfortunately, high-quality aspheric lenses are very expensive to produce. Molded plastic and glass lenses are easy enough to manufacture, but their quality is low. Better, though much more expensive, aspheric lenses can be shaped by diamond turning with a computer-controlled lathe. Another route, advanced by Tamron, involves applying optical resin that can be shaped to the surface of a lens. Even with high manufacturing expenses, many modern lenses make use of one or more aspheric elements. For example, the Canon 17–40 mm and 10–22 mm wide-angle zoom lenses each have three aspheric lens elements (see [Canon 03]).

Color fringes, or chromatic aberration (CA), require some discussion of optical materials. The problem is that the refractive index of glass or any transparent material depends on the wavelength of light, a property known as dispersion. Figure 8.3 illustrates the effect of dispersion on light rays by lens elements. The refractive index for blue light is higher than for red light, so blue rays are bent through larger angles as shown in the figure.

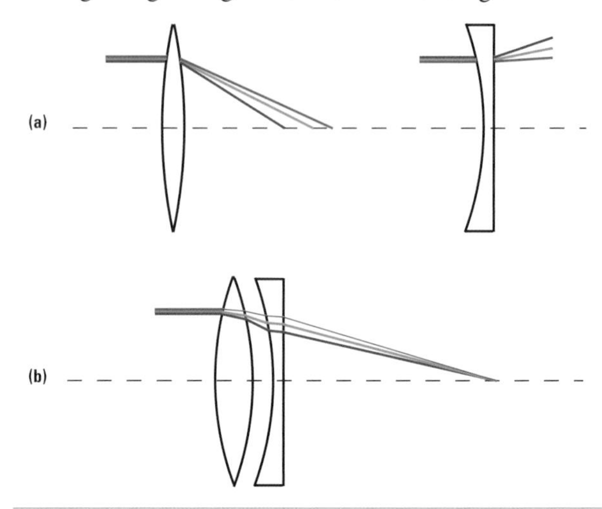

FIGURE 8.3. (a) The refraction of light by positive and negative lenses; (b) the combination of a positive lens with a negative lens having a higher index of refraction to produce an achromatic compound lens.

One way to compensate for dispersion is to combine lens elements (Figure 8.3(b)) that have been carefully selected to cancel chromatic aberration at two well-separated wavelengths. The positive lens element is often crown glass (made from alkali-lime silicates) with a relatively low refractive index, and the negative element is flint glass (silica containing titanium dioxide or zirconium dioxide additives) with a higher refractive index. These elements can be cemented together to make a single achromatic lens that has much less variation of focal length over the selected wavelength range than is possible with the a single glass element.

An *achromat* is a lens that has been corrected so that it has the same focal length at two wavelengths, usually in the blue and red regions of the spectrum. Lenses that have been corrected, perhaps with three elements, to give identical focal lengths at three wavelengths and to provide some spherical correction as well, are called *apochromats*. The best color-corrected lens, however, is the *superachromat,* which gives perfect focus simultaneously at four wavelengths of light and often into the infrared region as well. This near-perfect correction requires the use of expensive low-dispersion optical materials.

8.3 The Petzval Sum

The great lens designer Rudolf Kingslake said, "The design of any photographic lens is dominated by a certain mathematical expression known as the Petzval sum." He was referring to the specification of the image curvature that is produced by any simple or compound lens. This curvature, shown in Figure 8.4, is such an obvious characteristic of a simple lens that some opticians do not even regard it as an aberration.

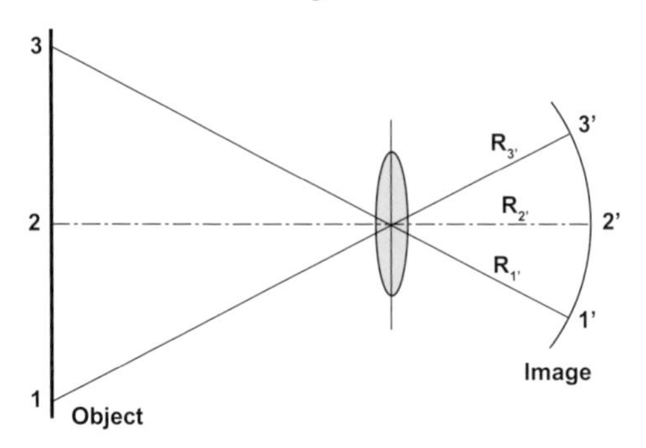

FIGURE 8.4. An illustration of image curvature.

This curvature is, in fact, a unique kind of aberration because it does not depend on the aperture, or other characteristic distances in compound lenses. It does, however, depend on the focusing powers of the optical surfaces, and in 1839 Joseph Petzval (1807–1891) derived an equation for the field curvature expected from a compound lens with a number of optical surfaces in the absence of other aberrations. He used the fact that the inverse of the image curvature ($1/R_p$) resulting from the ith surface of an optical element is proportional to the focusing power of the surface defined as $P_i = (n_i - n_{i-1}) / R_i$, where n_i and n_{i-1} are the refractive index before and after the surface and R_i is the radius of curvature of the surface. Petzval showed that the curvature for the image plane resulting from all the lens surfaces is given by the summation

$$(1/R_p) = n_\alpha \sum_i P_i / (n_i n_{i-1}),$$

where n_α is the refractive index of the last medium. When one says that the Petzval sum for a lens is large, they just saying that the image curvature is large. Of course, the aim is to make the Petzval sum as close to zero as possible by a combination of positive and negative surface powers.

8.4 Optical Materials

A good material for a lens must not absorb light in the visible region, but it also must have a refractive index greater than 1—and the higher the better. Of course, all materials absorb some light; and, if the thickness is great enough, the absorption will be evident. The normal situation in optics is for the refractive index to increase as the wavelength of light decreases so that the index of blue light is greater than that of red light. This effect was modeled by the mathematician Augustine Cauchy in 1836, and in 1871 his empirical equation was improved by W. Sellmeier to obtain the equation still used to characterize optical materials.[2] This equation relates the refractive index to the wavelength by means of empirically determined constants:

$$n(\lambda) = \sqrt{1 + \frac{B_1\lambda^2}{\lambda^2 - C_1} + \frac{B_2\lambda^2}{\lambda^2 - C_2} + \frac{B_3\lambda^2}{\lambda^2 - C_3}} \quad (8.1)$$

where $n(\lambda)$ indicates that n depends on the wavelength of light λ, and the B and C coefficients are used to describe optical glasses in the standard (Schott) catalog of glasses.

Optics catalogs also characterize the dispersion of materials by means of a standard parameter known as Abbe's number, v_D (see [Melles Griot 09a; Melles Griot 09b; Carl Zeiss 05; Nave 09]). This parameter is widely used, although it is a rather crude measure of dispersion. The definition is

$$v_D = \frac{(n_D - 1)}{(n_F - n_C)} \quad (8.2)$$

where n_D, n_F, and n_C are the indices of refraction at the wavelengths of the Fraunhofer lines helium D (yellow, 589.2 nm), hydrogen F (blue, 486.1 nm), and hydrogen C (red, 656.3 nm), respectively, that are found in the spectrum of the sun. In some treatments, mercury (Hg) lines and cadmium (Cd) lines are substituted, but the wavelengths are not much different. In the definition of v_D, the numerator is a measure of the refractivity while the denominator is proportional to the dispersion.[3]

The most commonly used optical glasses are crown ($v_D > 50$, $n \approx 1.5$) and flint ($v_D < 50$, $n \approx 1.7$). The problem is that the dispersion tends to be large for glasses that have large refractivities. This is evident from the Sellmeier equation. Absorption wavelengths are usually in the UV part of the spectrum, and, when an absorption wavelength is close to the visible region, both the value of the refractive index and the rate of change of the index with wavelength tend to be large. The Schott glass company has published plots of refractive index versus the Abbe number so that lens designers can select appropriate glasses. Basically, one can easily get high refractivity or low dispersion, but not both.

As with the problem of spherical lens shapes, one must live with the limitations imposed by nature. As previously noted, one way to minimize color fringing (chromatic aberration) is to play off different kinds of glass against each other. Another possibility is to search for natural and synthetic materials that push the limits of the Sellmeier equation by providing low dispersion with acceptable refractive indices. This could result from a fortunate set of absorption frequencies. For example, crystals of fluorite (CaF_2), which has $v_D = 95$ and $n_D = 1.43$, occur naturally but can also be grown for use in lenses. Ernst Abbe was the first person to use fluorite to enhance chromatic correction in microscope lenses in the 19th century. Fluorite is difficult to work with, and its refractive index is so low that lens elements must be thick and heavy. So lensmakers have turned to proprietary engineered glasses that are claimed to have low dispersion and fairly high refractive indices. For example, we find Hi-UD (Canon), ED (Nikon), AD and LD (Tamron), and SLD (Sigma) low-dispersion glasses advertised for new lenses.

A discussion of corrections for chromatic aberration would not be complete without mentioning diffractive optics. The diffraction effect from a concentric lattice of transparent plates, as in a Fresnel lens, disperses red light more than blue—exactly the opposite of a glass prism or lens. The idea is to use both diffractive and dispersive lens elements in a compound lens to compensate for chromatic and possibly other aberrations. Thus a lens element is introduced that has a diffraction grating etched into it. Early work in this area was delayed by problems with excessive diffraction-based flare, but now Canon has developed multilayer diffraction optics that are competitive at least for telephoto lenses. The primary advantage appears to be more compact and lighter lenses. For example, the Canon 400 mm *f*/4 DO IS lens is claimed to be reduced 37% in length and 31% in weight from a comparable lens designed only with refractive optical elements. At this time, only Canon offers lenses incorporating diffractive optics, and the prices are very high. Nikon does, however, offer a teleconverter lens (TC-E3PF) based on diffraction optics.

8.5 Anti-Reflective Coatings

It turns out that, with current technology, compound lenses incorporating many elements are required to correct various aberrations, and commercially available modern lenses bear witness to this requirement. Samples from lens catalogs show lenses with large numbers of elements. Also, low-dispersion glass and aspheric elements are used in crucial positions to improve performance. For example,

- Canon EF 100–400 mm *f*/4.5–5.6 IS USM, 17 elements in 14 groups (fluorite and Super UD-glass elements)

- Sigma APO 80–400 mm $f/4.5$–5.6 EX OS, 20 elements in 14 groups (2 SLD elements)
- Nikon 80–400 mm $f/4.5$–5.6 ED AF VR Zoom Nikkor, 17 elements in 11 groups (3 ED elements)

The modern multi-element lens is only possible because of a technological breakthrough that we have not yet discussed. In the 19th century, lenses with more than four air/glass surfaces suffered from low contrast. Some lenses with six glass/air surfaces were considered acceptable, but the eight air/glass surface lenses, while better-corrected for aberrations, were found to yield negatives with very low contrast. The problem was the reflection of light at the surfaces. When light passes from a region with the refractive index n_1 to a region with the refractive index n_2, a fraction R is reflected. Equations for R were worked out in the early 19th century by Augustin-Jean Fresnel. In the limit that the incident light is normal to the surface, he found that

$$r = \left[\frac{n_2 - n_1}{n_2 + n_1}\right], \quad R = r^2 . \tag{8.3}$$

Here r and R are associated with the amplitude and intensity of the reflected light, respectively. (The amplitude r refers to the electric field of light, and it will be needed to explain the effect of lens coatings.) For example, at an air/glass interface with $n_1 = 1$ and $n_2 = 1.5$, the fraction of reflected intensity is $R = 0.04$, or about 4%. This fraction is lost at every air/glass surface, so with four surfaces about 15% is lost and only about 85% of the incident intensity is transmitted. This was, indeed, a serious and disabling problem for lens technology. Light was not only lost, but the reflected light bounced around inside the lens, producing ghosts, flare, and a general loss of contrast in the focused image.

The solution to the problem lies in sophisticated anti-reflection coatings. Everyone with a camera has, of course, observed the beautiful violet and pastel shades that are reflected by several elements in the lens. The reflections from lenses manufactured in different time periods are shown in Figure 8.5.

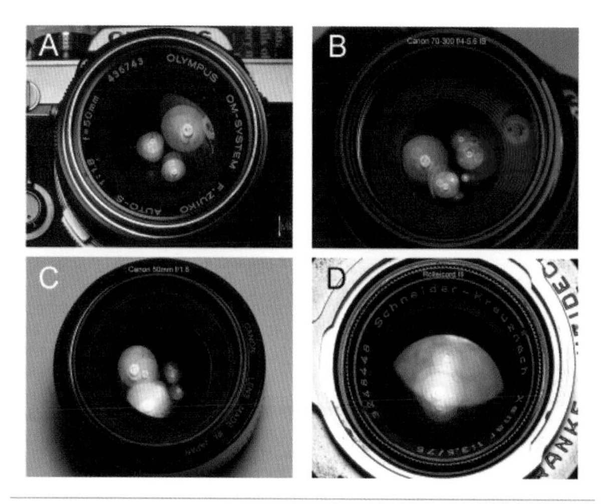

FIGURE 8.5. Reflections of a 5000 K lamp in various lenses. (a) Olympus 50 mm F/1.8 (1972), (b) Canon 70–300 mm F/4–5.6 (2006), (c) Canon 50 mm F/1.8 (1987), (d) Schneider-Kreuznach 75 mm F/3.5 (1953).

The colors reveal coatings that are themselves colorless but that increase the transmission of a broad range of frequencies in the visible spectrum while reflecting a small amount of light that has not been completely cancelled by interference. The discovery of the effects of lens coatings has an interesting history. Isaac Newton observed the "colours of thin transparent bodies" c. 1672–1675 and reported his finding in the second book of *Opticks* (1704). He knew about missing reflections associated with half-wavelengths, but his ideas are very hard to follow, and there were no practical applications. In the 19th century, with the invention of photography and advances in optics, the time was right for anti-reflective optical coatings. The early days of optical coating have been reviewed in a delightful article by Angus Macleod [Macleod 99]. Basically, the beneficial anti-reflecting property of coatings was accidentally discovered several times by famous scientists. For example, in the early 19th century, Joseph von Fraunhofer observed the reduction of reflection from glass that had been tarnished by acid treatment. Lord Rayleigh (1886) noted that old lenses (aged glass) transmitted light better than new (clean) lenses, and suggested the presence of a surface layer with a lower refractive index than the glass. The idea that "age coated" as well as intentionally tarnished lenses were preferable to those with bare air/glass surfaces was floating around by the 1890s. It appears that the first person to attempt the commercial use of coated glass was Dennis Taylor, who patented the use of various chemicals to tarnish optical surfaces deliberately in 1904. Unfortunately, Taylor's results were inconsistent, and coating development lagged until the 1930s.

The next breakthrough occurred in 1935, when Alexander Smakula at the Zeiss company developed a method for coating lens surfaces in a vacuum with an evaporated layer of low-index material. At about that time optics research became part of the war effort, and developments were classified. By 1941 the transmission of light by multi-element lenses had been improved by about 60%, and anti-reflection coatings were being ordered for periscope and binocular lenses. Macleod notes that coated binoculars permitted an extra 30–45 minutes of viewing time at dusk and dawn. Also, multi-layer coatings were being developed in Germany, and by the end of the war optical coating was a big business.

Since World War II, multi-layer coatings have been developed both to enhance and to diminish reflection, to select wavelengths for transmission or reflection, to polarize light, to split beams, and so on. A major driving force for the improvement of optical coating technology has been the development of lasers. Gas lasers, especially, e.g., the He–Ne laser, require highly reflective multi-layer coatings for mirrors. Of course, photographers have benefited greatly from the availability of multilayer-coated lenses. To quote Rudolf Kingslake, "The discovery of antireflection coating of lenses opened up the whole field of multi-element optical systems, so that now almost any desired number of air-spaced elements can be used without danger of ghosts and flare. Without such coating all of the modern zoom and other complex lenses would be impossible."

Now for the science. How do anti-reflection coatings work [Heavens 91]? Suppose a thin transparent coat with a refractive index of $n_1 = 1.25$ is applied to a glass lens that has the refractive index $n_2 = 1.5$. According to Equation 8.3, the fraction of intensity reflected at the air/coat surface is 1.2%, and the fraction reflected at the coat/glass surface is 0.83%. The important question is how to obtain the total reflection, and the numbers we have calculated are not helpful at all! Experiments show that the total reflection depends critically on the thickness of the layer. Transmission and reflection from thin layers are interference phenomena, and amplitudes of the electric fields must be combined before squaring. The amplitude of a reflected wave, which should be thought of as the length of an arrow (vector), is given by r (not R) in Equation 8.3 and each arrow has an orientation. This idea is discussed in Appendix B. A single-layer anti-reflective coating is shown in Figure 8.6. (The arrows in this figure indicate the directions of light rays and the associated amplitude arrows denoted by the r's are different things.)

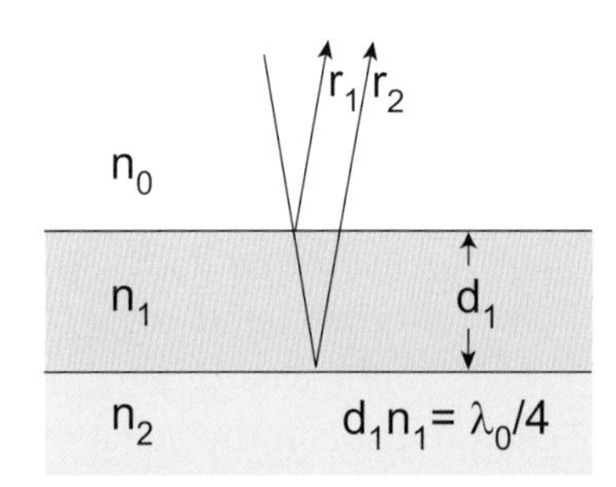

FIGURE 8.6. Single-layer anti-reflective coating with exaggerated angle of incidence.

Suppose that $n_0 = 1$ (air) and $n_2 = 1.5$ (glass). The coating layer must have a refractive index less than 1.5 (for there to be the same change in phase angle or turn of the amplitude arrow at the two surfaces), and, for effective cancellation of reflection, it is necessary that $r_1 = r_2$. This condition can be combined with the equation for r to show that the optimum refractive index for the layer is $n_1 = \sqrt{n_0 n_2}$, or in this case 1.22. We have a situation in which there are two ways for a photon to get from the source to the observer, and, according to the ideas in Appendix B, each of the paths can be represented by an arrow with a length and an angle. These arrows must be determined, added together, and the resultant arrow length squared in order to determine the intensity of the reflected light.

The lengths of the arrows are essentially equal because of the choice of n_1, and the difference in orientation (angle) depends only on the difference in the optical path lengths. The orientation of the arrow associated with r_1 (path 1) is arbitrarily set at 12 o'clock, and the orientation of the arrow for path 2 is different only because of the time required for the photon to traverse the layer of thickness d_1 twice with the effective speed c/n_1. A complete rotation of the arrow through 360° occurs each time the path length increases by one wavelength. With our choice of layer thickness, the arrow for path 2 is oriented at 6 o'clock, or 180° out of phase with the arrow for path 1. The two arrows, when placed head-to-tail, add to 0, and there is complete cancellation of the reflected light at wavelength λ_0. Since the sum of the intensities of the reflected and the transmitted light must equal the intensity of the incident light, this coating permits 100% of light at λ_0 to be transmitted. (We have simplified things here by assuming that the angle of incidence is close to zero, the amplitudes of reflection are very small, and by neglecting multiple reflections in the layer. Corrections for these simplifications are well known and do not change the qualitative picture.)

Of course, things are not perfect, because the single layer is only efficient at canceling reflection at one wavelength. With a two-layer coating it is possible to zero out reflection at two wavelengths, a three-layer coating can be designed to cancel reflection at three wavelengths, etc. Approximate analyses of these situations are quite easy with the vector method summarized here.

8.6 Conclusion

We set out to show why modern lenses are so complicated. That should be evident by this point. But it should be noted that the discussion thus far applies to fairly simple "prime" or non-zoom lenses. Also, most modern lenses contain auto-focus mechanisms and in some cases vibration-isolation systems based on the movement of a lens element to counteract low-frequency vibrations. These important technological refinements are beyond the scope of this book, but some of the optical characteristics of compound lenses, e.g., primary surfaces and nodal points, will be considered in later chapters.

Further Reading

R. Kingslake. *Lenses in Photography: The Practical Guide to Optics for Photographers.* Rochester, NY: Case-Hoyt Corp., 1951. (See Chapter 4).

CHAPTER 9

Coming to Terms with Real Camera Lenses

Any sufficiently advanced technology is indistinguishable from magic.

—ARTHUR C. CLARKE

9.1 Introduction

We have seen why compound, multicoated photographic lenses are needed to produce aberration-free images with high contrast, but we have not considered how to deal with lenses that often look like pipes with glass at each end. The overall length of camera lenses is often greater than their effective focal length, and it is not obvious which parameters are significant. Indeed, some extreme telephoto lenses are truly "monster lenses." For example, the Canon 400 mm, $f/2.8$ lens is 13.7 in. long and weighs 11.8 lbs, and it is not even the heaviest lens available for a 35 mm (FF) camera. It is natural to ask how we can get away with discussing focal length, field of view, depth of field, etc., in terms of single-element lenses when real lenses look nothing like that. Fortunately, simple lens ideas and equations are still useful and provide good approximations to compound-lens behavior (as long as objects are not too close to the lens), but there is a lot more to understand.

Clearly, photographers do not need to know all the internal details of the lenses they use, but it is satisfying to understand the primary characteristics of the beautiful modern lenses that are available. Also, there are situations where a user needs information about lens parameters; for example, in dealing with panoramic pictures constructed by stitching together overlapping images, which often do not match properly for nearby objects. The problem is that proper alignment of objects in the images requires the camera to be rotated around the proper axis. This "no-parallax axis" must be determined either by doing experiments or by visually locating the *entrance pupil* of the lens as discussed in this chapter. (This assumes that the sensor rotates with the lens.) The goal of this chapter is to explain how camera lenses behave without a detailed consideration of all the internal components. For the most part, we can get by with a "black box" approach.

Let's begin by trying to imagine a perfect simple lens. This is not as straightforward an exercise as one might think. If technology and materials pose no limitations, what do we want in a lens? First, rays of light from points on an object must pass through the front surface of the lens and then emerge from the rear surface directed at precisely the correct points in the image plane. All of the rays from a properly focused object point must reach the image point at exactly the same time so that they are in phase and contribute to the intensity. Furthermore, perfect focus must occur for all colors in order to form an aberration-free image. This perfect lens has aspheric surfaces and is fabricated from a wonder material that has a high refractive index but no dispersion. But wait—modern lenses, while not perfect, do more than that. We expect variable focal length (zoom), automated focus and aperture, and even image stabilization. We can get most of those features out of the way for the perfect lens by assuming that the material properties, e.g., the value and spatial distribution of the refractive index, can be controlled externally by electric and magnetic fields. For the present purposes, I concentrate on static optical properties, but I note that the surfaces do not have to have any particular shape since the optical path length can, in principle, be controlled with internal variation in the refractive index.

So the perfect lens is nothing more than two surfaces with something between them that transforms light rays incident on the front surface into rays that exit the rear surface with the correct phases and directions. With current technology, we can only approximate the optical properties of the perfect lens. That is to say, our real camera lens will have some kind of entrance and exit surfaces and between them will be a "magic zone" or black box, containing as many lens elements with various refractive indices as are necessary to transform the incident rays into the desired exit rays. This picture of real lenses is a good starting point for relating simple lenses to compound lenses, and I begin by reviewing the way simple lenses behave.

9.2 The Thin Lens

A *thin lens*, as shown in Figure 9.1, has an axial thickness that is much less than the lens diameter and the focal length, and distances along the axis can be measured from the central plane of the lens. Rays, from objects to the left, that are parallel to the optical axis, after refraction, are focused at F_2, the rear principal focus point. For objects at less than infinite distances, a real image will always be formed at a distance greater than f from the lens. The imaging properties of this lens can be specified with a short set of rules:

1. Rays passing through the center of a thin lens are undeviated.
2. Rays parallel to the optical axis pass, after refraction, through the focal point (F_2) of the lens.

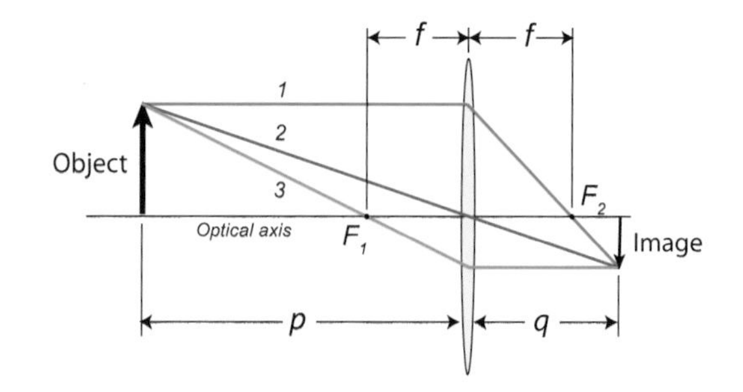

FIGURE 9.1. Image creation with a simple lens.

Of course, this works in both directions, so a ray passing through a focal point will, after refraction, emerge parallel to the optical axis. These rules suffice for determining the magnification m = image/object = q/p and also for deriving the conjugate equation $1/p + 1/q = 1/f$. With this lens there is no ambiguity. The distances are well defined, and, if a stop (aperture) is added, it will certainly be very close to the front or rear surface of the lens.

9.3 The Compound Lens

For the compound lens we begin with an approximate picture based on first-order optics. The thick lens shown in Figure 9.2, with its cardinal points (principal and nodal) and front and rear focal points is an idealization of a real compound lens. It is accurate for paraxial rays, i.e., rays near the optical axis, and making small angles with that axis.

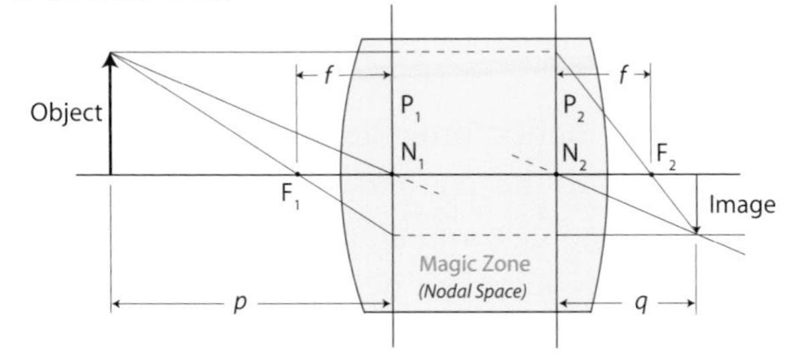

FIGURE 9.2. Image formation by a compound lens.

The vertical lines P_1 and P_2 represent the first and second principal planes, and N_1 and N_2 are the corresponding nodal points. Assuming that the refractive indices are the same in the object and image spaces, the nodal points lie in the principal planes and the front and rear focal lengths, f, are equal. We do not have to specify what occupies the Nodal Zone or "hiatus," but only the nature of the transformation it produces. The rays do not follow the extensions shown in the nodal space, but act as if all deviations occur at the principal planes. We can describe the behavior of the compound lens with a short (incomplete) list of rules describing its effect on incident rays lying in a plane containing the optical axis (meridional rays):

1. Rays passing through P_1 parallel to the axis also pass through the focal point F_2 on the opposite side.
2. A ray passing through N_1 from the left emerges from N_2 at the same angle with the axis.
3. A ray entering P_1 at a height h above the axis exits P_2 at the same height from the axis.

This thick lens behaves as though there are two refracting surfaces, and the distance from the principal plane P_2 to the focal point is called the focal length. The conjugate equation can also be carried over from the simple lens by measuring p and q from the principal planes rather than from the center of the lens. Also, the field of view shown in Figure 9.3 can be calculated using the simple lens equation $FOV = \theta = \tan^{-1}(S/2f)$. Here S is a sensor dimension and can be defined as the width (horizontal), height (vertical), or diagonal.

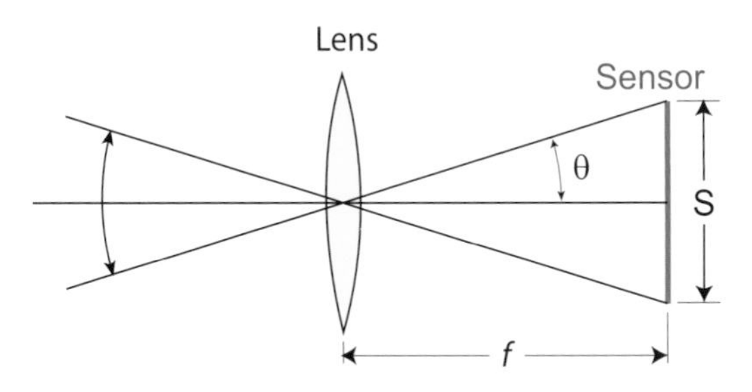

FIGURE 9.3. The field of view for a simple lens.

Before becoming too comfortable with the thick lens as shown, however, one must realize that the principal planes are mathematical constructions. They can, in principle, have any separation and, in fact, P_2 can be located in front of P_1.

9.4 Gaussian Optics

So how can this simple picture based on the cardinal points be justified? What we need to show is that the equations that describe a simple thin image-forming lens also describe the behavior of a compound lens. To do this, we turn to the theory known as Gaussian optics that is attributed to Carl Friedrich Gauss (1777–1855).[1] This theory applies to light rays that remain close to the optical axis. (The angles with the axis of such rays are small, and when they intersect an optical surface they make a small angle with the surface normal. This is exactly the condition for first-order optics to hold where, for any angle θ, $\sin\theta$ can be replaced with θ itself.) The compound lens, no matter how complicated, consists of spherical surfaces separating regions with constant refractive indices. The "power" of a surface is given by $(n_1' - n_1)/R$, which is just the change in the index of refraction on passing through the surface divided by the radius of the surface. In a meridional plane, i.e., a plane that includes the optical axis, each optical surface appears as a segment of a circle with its center on the axis.

If the math is getting scary, hang on, because the ideas are really very simple. The situation is shown in Figure 9.4. At any time, a ray is described by just two things: its distance (x) from the axis and the angle (α) it makes with the axis.

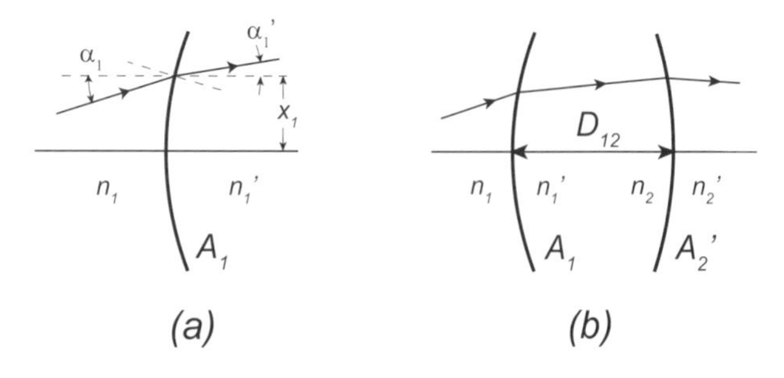

FIGURE 9.4. (a) A ray passes through surface A and the angle changes from θ_1 to θ'_1; (b) a ray passes through a simple lens consisting of two surfaces, A_1 and A_2, and the space between them.

As the ray passes through a compound lens, it can be transformed in two ways:

1. The ray passes through a surface, and its angle with the axis changes but not its distance from the axis.
2. The ray translates, and its distance from the axis changes but not its angle with the axis.

The effect of a single surface is shown in Figure 9.4(a), where the angle changes but the height does not. The combined effect of a surface, a space, and another surface is shown in Figure 9.4(b). The effect of a surface depends on its power, while the effect of a translation depends only on the angle of the ray and the distance traversed.

In short, here is the story: One can work out, in a straightforward way, the effect on a light ray of all the optical surfaces and the spaces from the first surface to the last surface. The mathematical description of this transformation can then be compared with the transformation produced by a single thin lens. It turns out that the two descriptions are identical in form if and only if the computation for the compound lens begins and ends at the proper principal planes. Locating these planes is also straightforward, but there is no requirement that they be located close to real surfaces and, as noted above, the second principal plane may appear to the left of the first principal plane. The complete details can be found in standard optics texts. Also, an outline of the theory is given in Appendix D.

Gaussian optics provides a good description for rays close to the axis and can provide a starting point for designing compound lenses, but, of course, ray tracing is required for designing modern photographic lenses. One place that Gaussian optics is still very useful is in setting up systems of lenses to direct laser beams. The beams of nearly parallel light rays, i.e. collimated beams, are usually close to the axis and furthermore are monochromatic.

9.5 Entrance Pupil, Stop, and Illumination of the Image

At this point, we want to consider the factors that determine the illumination of the image (see [Jacobson 09]). The brightness of the object is the first factor. Light from an object O is radiated into the cone defined by the half-angle θ, and a fraction of that light is transmitted by the optical system to the image I. The image is illuminated by a cone of rays defined by the half-angle θ' as shown in Figure 9.5.

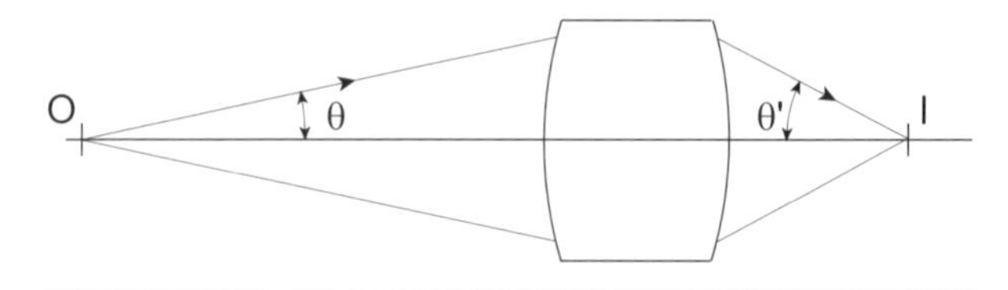

FIGURE 9.5. The object O, the optical system, and the illuminated image I.

The important fact is that the illumination of the image is proportional to the brightness B of the object and the square of sin θ', i.e., the illumination is proportional to $B(\sin\theta')^2$. Figure 9.6 illustrates the image side of a compound lens and correctly shows that second principal "plane" in a corrected lens is, in fact, a refracting spherical surface.

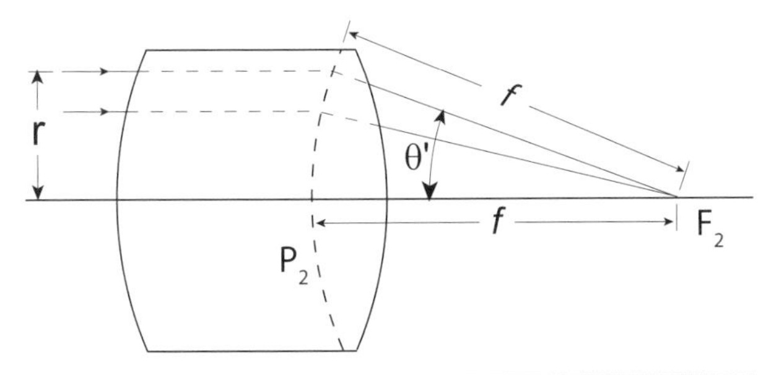

FIGURE 9.6. The spherical refracting surface and the cone of light illuminating an image at F_2.

When distant objects are photographed, it is appropriate to replace sin θ' with r/f. This substitution motivates the definition of the F-number of the lens:

$$\text{F-number} = N = \frac{1}{2\sin\theta'} = \frac{f}{2r} .$$

Here the diameter $2r$ refers to the effective aperture of the lens. This is not the diameter of the front lens element or even the iris diaphragm. The *effective aperture* or *entrance pupil* is the image of the aperture stop as seen from the front of the lens. What we mean by "stop" is a physical aperture that limits the cross-section of the image forming pencil of light. The stop is commonly an opaque sheet with an adjustable (iris) aperture, but it may be the edge of a lens or a diaphragm. The aperture stop can be anywhere in the compound lens or even outside the lens. In lenses that have been corrected for spherical aberration and coma, the equivalent refracting surface is part of a sphere. This sphere is centered on the focal point, and a simple construction as in Figure 9.6 shows that the radius r can never exceed f. The conclusion is that the theoretical maximum aperture corresponds to $N = 0.5$.

The important point is that we see the effective aperture when we look through a lens from the front. It is this apparent diameter that must be used for computing the F-number and the depth of field. Figure 9.7 shows the effective apertures for two lenses.

FIGURE 9.7. The entrance pupils of two lenses.

The entrance pupil of the lens on the left is very close to the front of the lens, while the entrance pupil for the lens on the right is much farther back. Also, the apparent location of the effective aperture moves as the focal length of the zoom lens is changed. It is the apparent location of the stop that is the pivot point or "no-parallax" axis of the lens (see [Littlefield 09]). The entrance pupil is also the perspective point of the lens. Fortunately, some manufacturers publish the nodal points and pupil points for their fixed focal length lenses online (see, for example, the nodal point and pupil point table for Sigma fixed-focus lenses at http://www.sigma-foto.de/cms/upload/downloads/informationen/Nodel_point_fixed.pdf).

9.6 Vignetting

The loss of intensity at the edges and particularly the corners of an image is known as vignetting (see ["Vignetting" 09]). This effect can be desirable for removing distractions and drawing attention to the center of a photograph, but unintentional vignetting is a defect that needs to be minimized or removed. There are at least three sources of vignetting. The most obvious one results from the obstruction of oblique rays by features in the lens barrel or lens attachments such as filters or hoods. I refer to both internal and external "cut-offs" as *mechanical vignetting*, although some authors prefer *optical vignetting* for the effects of internal obstructions. From Figure 9.7 it is clear that internal blockage will be most pronounced with large apertures and long lenses. An extreme example of mechanical vignetting resulting from the use of an inappropriate lens hood on a 28 mm lens is shown in Figure 9.8. External obstructions can, of course, be removed, but we are stuck with the obstructions resulting from lens design.

FIGURE 9.8. Example of mechanical vignetting caused by a lens hood.

The variation of illumination of images over the field of view even in the absence of mechanical vignetting also exists. This *natural vignetting* is geometric in origin and is approximated by the $\cos^4$ law. The following is a simplified derivation of the $\cos^4$ law based on Figure 9.9. (It is instructive and reasonably good for small aperture lenses.) The illuminance of an area on the sensor at angle ϕ is reduced relative to the illuminance at angle zero by four factors of $\cos\phi$ (as defined in Figure 9.9).

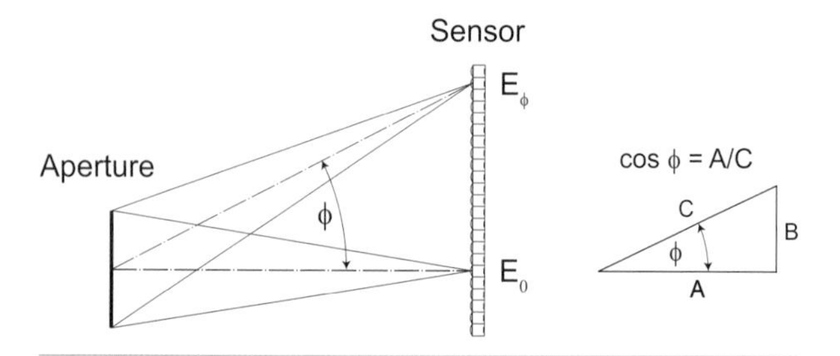

FIGURE 9.9. Geometric factors determining illumination of detectors in sensor.

The first $\cos\phi$ factor results from the oblique angle at which the circular aperture is viewed. For the off-set detectors at angle ϕ, the aperture appears to be an ellipse with height and area reduced by a factor of $\cos\phi$. The second factor arises because a beam of light striking the sensor at φ is spread over a larger area of the sensor than a beam normal to the sensor. Finally, the illuminance is inversely proportional to the square of the distance from the aperture. The ratio of the distances to the sensor at angle ϕ and angle zero is C/A or $1/\cos\phi$ and the inverse of this quantity squared gives the required factor of $\cos^2\phi$. Thus we have found that the illuminance at angle ϕ is reduced by a factor of $\cos^4\phi$, and this result has been derived by using only the geometry on the image side of the lens. For some lens designs it is preferable to determine the first cosine factor from the object side, where the apparent shape of the entrance pupil also depends on the angle of the oblique ray. This approach will give a different angle in the first cosine factor if the lens has been specially designed to increase the aperture-to-sensor distance. Such *retrofocus* lenses are often found in DSLRs, where space must be available for the mirror between the lens and the sensor.

With digital cameras, there may be yet another mechanism for intensity loss at the edges of the image. As shown in Figure 9.9, the detectors do not lie on the surface of the sensor but rather at some depth in the chip, effectively down a "tunnel." In CMOS sensors they occupy the pixel site along with other circuit elements, and it is common for the pixel to be covered by a small lens that directs light down the optical tunnel to the photodiode. The problem is that oblique rays are not as efficiently captured by the pixel lens as are rays normal to the sensor. The resulting loss of intensity has been called *pixel vignetting*. The problem is less severe with CCD detectors where active transistors at the pixel site are not required for readout and the entire pixel can be light sensitive. In this case, the pixels are defined by thin gates.

CCD sensors have the advantage of efficient light collection, but the advantage is not significant for large pixels, i.e., 10 μm or greater.

It turns out that pixel vignetting is a more severe problem for digital versions of rangefinder-type cameras, e.g., the classic Leicas, even when CCD sensors are used. A rangefinder camera does not have a mirror box and does not need to use a retrofocus lens. Therefore, the lenses mount much closer to the sensor and the marginal rays strike the sensor at larger values of ϕ. Even with CCD sensors, the illumination of photodiodes in pixels at the edges is a challenge. The Leica M8 addresses this problem by, first, using a sensor that is not full frame (18 mm × 27 mm in place of 24 mm × 36 mm); second, using a sensor with pixel lenses off-set from the center of the pixels by varying amount to compensate for off-axis angles; and, third, correcting for the remaining intensity loss in software for each type of lens.

Software correction for vignetting is, of course, available in programs such as Photoshop. Photoshop provides a slider control for increasing or decreasing brightness in the corners and a midpoint slider for setting the area affected by the correction. Software correction is effective for all types of vignetting, including the fall-off of flash illumination away from the center that is common when wide-angle lenses are used.

9.7 Conclusion

Chapters 8 and 9 have provided an introduction to lens design issues but they are by no means definitive. The interested reader can find additional material on websites and, of course, in optics texts. Chapter 10 provides a brief look at fisheye lenses, Chapter 11 considers "equivalent images" obtained with lenses having different focal lengths, and Chapter 12 introduces lenses and combinations of lenses for close-up photography. We hope that the reader will gain some appreciation of the art, science, and technology of lens making and will be motivated to pursue interesting ideas elsewhere. In spite of its age, this field is very active! Every day new ideas and experiments are reported. For example, metamaterial-based superlenses are appearing that produce sub-diffraction-limited images (see [Smolyaninov et al. 07]). This is cutting-edge science, where negative refractive index materials are being produced by a variety of techniques.

Further Reading

R. E. Jacobson, S. F. Ray, G. G. Attridge, and N. R. Axford. *The Manual of Photography: Photographic and Digital Imaging.* Oxford: Focal Press, 2000. (See Chapters 4 and 5.)

R. Kingslake. *Lenses in Photography: The Practical Guide to Optics for Photographers.* Rochester, NY: Case-Hoyt Corp., 1951. (See Chapter 6.)

CHAPTER 10

Fisheye Lenses and How They Capture the Whole Sky

In this connection it is of interest to ascertain how the external world appears to a fish below the surface of smooth water. The objects surrounding or overhanging the pond must all appear within the circle of light . . . .

—ROBERT W. WOOD

This started me to look at the world in a different way. Ever since, sometimes I look with telephoto eyes, sometimes with wide-angle eyes.

—ALFRED EISENSTAEDT

10.1 Introduction

A number of severe problems are encountered in the design of wide-angle lenses with large apertures. First, there are the aberrations. Barrel distortion, where vertical and horizontal lines tend to bow out away from the center of the image, is common; and corrections become more difficult as the focal length decreases. For example, full-frame 35 mm cameras have an image diagonal of 43.3 mm, and the smallest practical focal length for highly corrected, large-aperture, rectilinear lenses turns out to be about 15 mm. But the real problems begin as the field of view advances toward 180°. In this range the width of the image increases without limit and the brightness at the edges vanishes. The decrease in brightness away from the center of the image results from the following effects, as shown in Figure 10.1.

The circular lens aperture is reduced to a narrow ellipse when viewed from large angles.

1. The circular lens aperture is reduced to a narrow ellipse when viewed from large angles.
2. At larger angles a bundle of light rays from the lens is spread over larger areas (s) of the sensor.
3. The distance (r) from the lens to the edges of the sensor becomes very large.

(These effects are combined in the $\cos^4$ law [Chapter 9], which states that the image illumination is proportional to the fourth power of the cosine of the angle of incidence.)

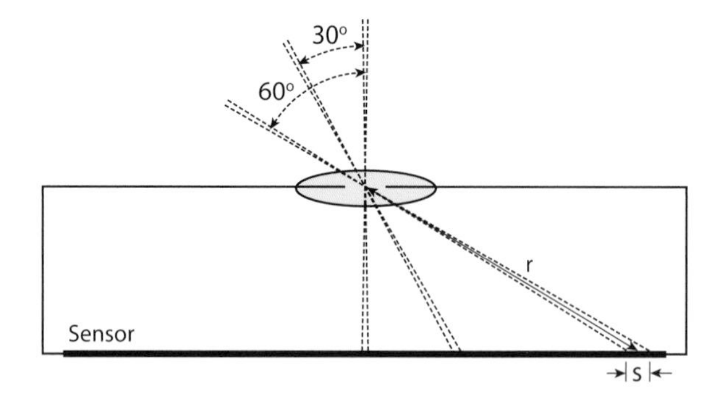

FIGURE 10.1. The geometry of light rays at different angles.

So how can we achieve wide angles of view even exceeding 180°? Fisheye lenses provide an answer if we are willing to accept a special kind of distortion or, rather, a different mapping of the three-dimensional world onto a two-dimensional plane. In fact, fisheyes can provide fields of view larger than 180°, and they are widely used in astronomy and meteorology (all-sky cameras), surgery (endoscopic imaging), surveillance (stationary CCD cameras), and personal security applications (peepholes). There is interesting science here with a long history. In striking contrast to rectilinear lenses, fisheye lenses yield circular images in which the hemispheric field is projected onto a plane. Very large distortions result from this projection.

Figure 10.2 provides a good comparison of the operation of a fisheye lens with a standard camera lens (rectilinear). This illustration is based on the classic *scioptric ball* or "ox-eye lens" that was developed by Daniel Schwenter (1585–1636) (see ["Daniel Schwenter" 09]). The lens is mounted in a ball so that it can be rotated to any orientation, much like the human eye. The beauty of this arrangement is that the bundle of rays close to the optic axis form an accurate image of part of the hemispheric field at each orientation.

With modern technology, this is sufficient, because the segments of the image can be captured and stitched together with a panoramic computer program. When the rays are projected on the planar sensor at the bottom of the figure, a well-corrected rectilinear image is obtained similar to the image obtained with a pinhole camera. Of course, the overall brightness is greater and the "swing lens" permits the light flux to be captured more efficiently at large angles.

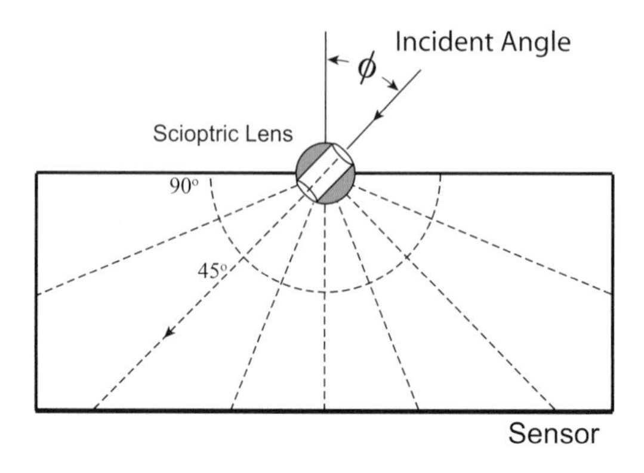

FIGURE 10.2. Image projection with a scioptric lens.

In contrast to this, a fisheye lens effectively captures the image on a hemispherical surface similar to that shown by the dotted curve. The question then becomes how to project the hemisphere onto a planar surface (see [Bourke 09]). There are, in fact, a number of different fisheye projections in use. Figure 10.3(a) shows an orthographic projection (OP) similar to the one obtained by the fisheye lens developed by Nikon for brightness measurements.

The *equidistant projection* fisheye shown in Figure 10.3(b) ensures that the distance of an image point from the center is proportional to the incident angle of the corresponding object point. Commercial fisheye lenses often approximate the equidistant projection. It is interesting to note that digital cameras are often rotated about their entrance pupils to acquire a set of images that can be used to construct a panorama. This is similar to imaging on a hemisphere with a scioptric lens. The stitching software that constructs the flat panorama is essentially generating part of an equidistant fisheye projection.

In comparing conventional wide-angle and fisheye lenses, we encounter the following projections

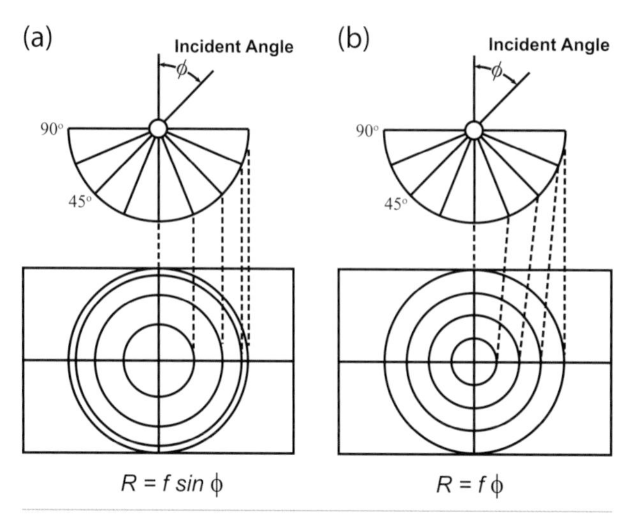

FIGURE 10.3. (a) Orthographic projection; (b) equidistant projection.

where R is the distance from the center of the image, f is the focal length, and ϕ is the incident angle.

10.2 Conventional or Rectilinear Lenses

Well-corrected rectilinear lenses provide the same projection as a pinhole camera. Parallel lines in planes perpendicular to the optic axis remain parallel in the image. The perspective is distorted at the edges of the image, and the field of view is limited by the focal length of the lens and the size of the sensor. For example, if $f = 15$ mm and the sensor has a width of 36 mm, the horizontal field of view (HFOV) is 100.40 ($R = f\tan(\phi)$.

10.3 Types of Fisheye Lenses

The *equidistant projection* fisheye (also known as *linear-scaled*, *angular*, or *ideal*): It is convenient for angular measurements, as in star maps. ($R = f\phi$).

The *orthographic projection* (OP) fisheye: Nikon was able to implement the OP fisheye by using aspheric lens elements. This lens gives an extreme fisheye effect with objects at the center made even larger and is useful for measuring illumination in architectural applications. This special-purpose lens is expensive and not very popular ($R = f\sin(\phi)$).

The *equisolid-angle projection* fisheye: With this lens, an element of area in the image is proportional to the corresponding element of solid angle in the scene. The mapping is similar to that obtained with a convex spherical mirror. One use of such lenses is to measure the fraction of sky covered by clouds. This type of projection is found in a number of commercial full-circle fisheye lenses for 35 mm cameras ($R = 2f\sin(\phi/2)$.

10.4 How Fisheye Lenses Accomplish Their Magic

The term *fish-eye view* was introduced into photography by Robert Wood, who was an optical physicist at Johns Hopkins University. In his text *Physical Optics* [Wood 11], he invited the reader to consider the outside world as viewed by a fish. Directly overhead objects would be seen clearly; but away from the vertical direction there would be distortion. Also, beyond a critical angle, the surface of the water would be a mirror because of total internal reflection. The effect can be quantitatively understood by considering refraction of rays of light at the air/water interface as illustrated in Figure 10.4.

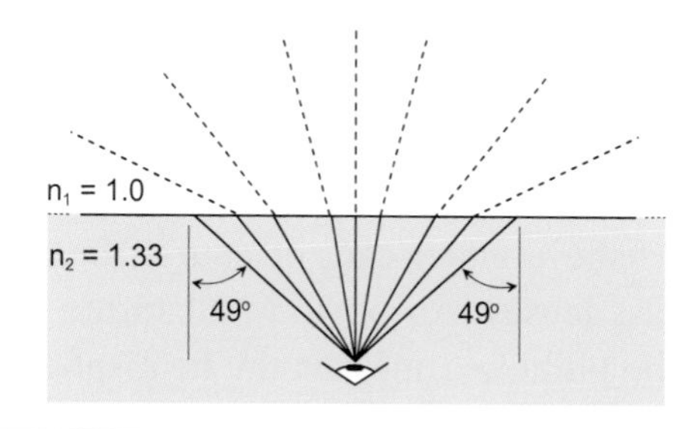

FIGURE 10.4. A fisheye view.

According to Snell's law, $n_1\sin(\theta_1) = n_2\sin(\theta_2)$, and since $n_1 = 1.0$ or air and $n_2 = 1.33$ for water, the angle of incidence θ_1 is always greater than the angle of refraction θ_2. (Refraction is illustrated in Figure 6.2.) Also, we find that when $\theta_1 = 90°$, the angle θ_2 of the refracted ray reaches a maximum value of about 49°. Figure 10.4 shows how the refracted rays in the water, restricted to angles less than 49°, map out the full 180°-view above water.

The human eye is not constructed to be able to see this effect very well, but Professor Wood recognized that it can easily be photographed. In fact, he constructed a pinhole camera, filled it with water, and demonstrated that it could produce photographs with the full 180° hemispheric view. Four of these photographs are reproduced in his text.

In Wood's pinhole water-tank camera, the distances from the center of the image are simply related to the angle of view with 0° at the center and 90° at the edge, as shown in Figure 10.5. The absolute distances, and hence the scale, depend on the distance from the pinhole to the film plate and the refractive index of the fluid in the tank.

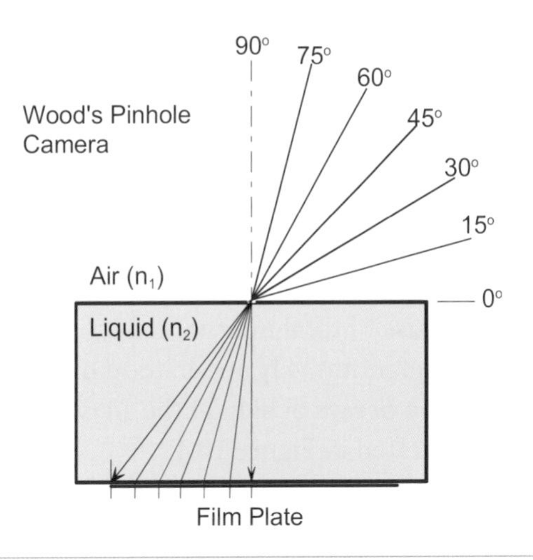

FIGURE 10.5. Illustration of Wood''s pinhole camera.

The fisheye effect can, of course, be realized with glass lenses. An obvious substitute for water in the pinhole camera is the hemispheric lens shown in Figure 10.6(a). With this lens, however, the Petzval sum is large, i.e., there is a lot of field curvature. The field curvature can be improved by adding a diverging (negative) lens as in Figure 10.6(b), but there are still significant color aberrations. Kenro Miyamoto [Miyamoto 64] showed that the color aberrations can be corrected by adding a doublet, with positive and negative elements, before the pupil (Figure 10.6(c)). Most modern fisheye lens share features of this design.

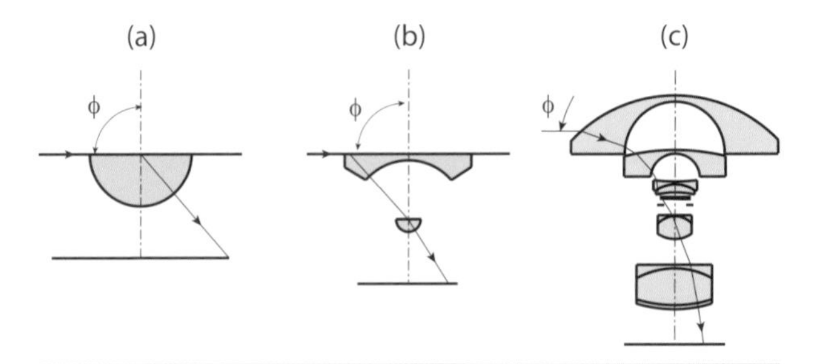

FIGURE 10.6. Evolution of fisheye lenses: (a) Bond's hemispheric lens, (b) Hill's fisheye, and (c) modern fisheye lens by Miyamoto.

The negative-entrance lenses divert incident rays so that the effective aperture is tilted. Also, the design decreases the exit angles and minimizes the differences in path lengths for rays in different parts of the image. The overall effect is to cancel the factors that contribute to the $\cos^4$ law for rectilinear lenses and to give much more uniform illumination over the image. Basically, the uncorrected barrel distortion distributes the light flux into small areas at large distances from the center of the image circle to give fairly uniform illumination.

10.5 Examples and Illustrations

Standard fisheye lenses for 35 mm cameras give 180° fields of view and are available in two types. Fisheye lenses with focal lengths in the range 6–8 mm fit the complete 180° field-of-view (full circle) disk into the narrow dimension of the image, while "full-frame" fisheye lenses with focal lengths of 14–16 mm spread the hemispherical image across the film or detector diagonal, e.g., 43.3 mm for FF sensors as shown in Figure 10.7. The most widely used DSLR cameras have sensors that are approximately 15 mm × 23 mm in size, i.e., with a crop factor of 1.6. With the smaller sensor, an equidistant fisheye lens with $f = 15$ mm will give a HFOV of only about 85°. Therefore, a lot of the fisheye effect is lost, and we are left with a wide-angle image with some barrel distortion. Of course, the barrel distortion can be corrected with image editing software such as Photoshop to provide a rectilinear image. A similar HFOV can be obtained with a conventional lens (rectilinear) having a focal length of about 11 mm. It is interesting to note, however, that the full-circle fisheye lens for a FF camera almost becomes a full-frame fisheye on a 1.6 crop-factor camera.

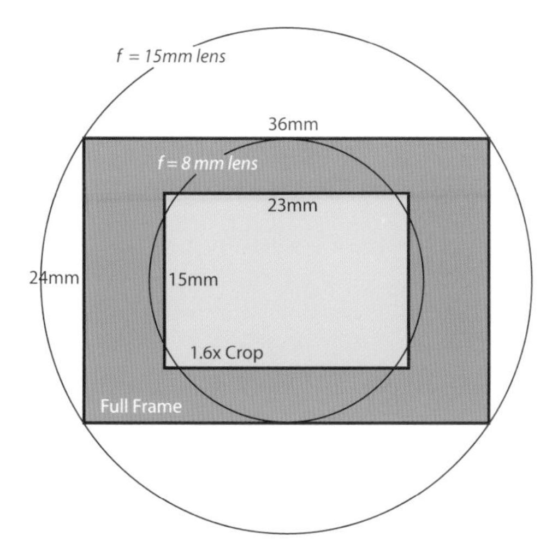

FIGURE 10.7. Image circles (180°) and sensor sizes.

As noted, computer correction of barrel distortion in fisheye lenses is now available. Consider the image in Figure 10.8, obtained with a full-frame (f = 15 mm) fisheye lens (Sigma EX) on 35 mm film. The yellow rectangle defines the area the image that would be obtained with a crop factor of 1.6.

FIGURE 10.8. Fisheye image of a scene at the North Carolina Arboretum in Asheville.

Notice that each small segment of the image has fairly small distortion even though the overall image shows strong barrel distortion. The effect is similar to that of looking through a transparent hemisphere divided into many sections, freezing the scene observed into each segment, and then projecting the hemisphere onto a plane. The apparent distortion results from the large angles displayed. By design, the angular spread from corner to corner diagonally is 180° and the horizontal and vertical fields of view are approximately 150° and 92° for FF sensors, respectively. Two horizontal lines will naturally appear to approach each other at the ends, as required by perspective. Now we convert the fisheye image to a rectilinear image by adjusting the distance of each image point from the center. The result of this conversion, as performed with Panorama Tools written by Helmut Dersch [Dersch 09], is shown in Figure 10.9 for full-frame and 1.6 crop-factor cameras. It is, in fact, the rectilinear image that conflicts with reality at very wide angles by attempting to counteract perspective. The distortions produced by ultra-wide rectilinear lenses are perhaps more disturbing than the fisheye distortions, and in both cases we are attempting to represent a wide view of the three-dimensional world with a planar image. It is analogous to the problems that mapmakers face in representing the globe with flat maps.

FIGURE 10.9. Illustration of the correction for barrel distortion in a fisheye image.

10.6 Conclusions

We have shown that fisheye lenses are not extensions of wide-angle lenses but, rather, a different optical concept. The full 180°-hemisphere can be imaged, and with available software the image can be converted to the rectilinear form if desired. Here we have a preview of the future of optics. If it is difficult to eliminate barrel distortion in a wide-angle lens, leave it in and correct the result with software. This and other in-camera corrections can be found in some modern digital cameras.

Further Reading

R. Kingslake. *A History of the Photographic Lens.* New York: Academic Press, 1989. (See Chapter 10, pages 145–149.)

"Fisheye Lens." *Wikipedia.* Available at http://en.wikipedia.org/wiki/Fishey_lens, 2009.

CHAPTER 11

What is an Equivalent Image?

Distinctions drawn by the mind are not necessarily equivalent to distinctions in reality.

—THOMAS AQUINAS

Size matters not. Look at me. Judge me by my size, do you? Hmm? Hmm. And well you should not. For my ally is the Force, and a powerful ally it is.

—YODA

11.1 Introduction

The advertisements for digital cameras usually state that the lens has a focal length range "equivalent" to a certain range on a 35 mm camera. One "expert" tells us each lens has a focal length and that it is not equivalent to any other focal length regardless of the circumstances. Another expert claims that a big sensor is necessary to get a small depth of field. These are at best half-truths. So, what is correct? In earlier chapters I explained the relevant background material concerning perspective, field of view, brightness, and diffraction. Here I just need to bring all that together. This exercise provides a good review of concepts as well as leading to useful conclusions. In particular, I show that increasing the F-number increases the depth of field, but also reduces the resolution through the effects of diffraction. The conclusion is that the sensor size alone determines the maximum useful F-number once the resolution has been specified.

It is easier to understand the important points by considering what is necessary to obtain an "equivalent" or identical image with different camera/sensor sizes. In order to obtain an equivalent image all of the following properties must be the same:

- Angle of view (field of view)
- Perspective
- Depth of field (DoF)
- Diffraction broadening
- Shutter speed/exposure

According to this definition, two equivalent images will appear to be the same in all respects; and the observer will not be able to detect any evidence of the size camera/sensor that was used. This still leaves "noise," pixel count, and lens quality to be considered later. The starting point is to recognize that typical cameras have sensors that range in size from about 4 mm × 6 mm to 24 mm × 36 mm, covering a factor of at least 6 in. width. Let's call the small sensor point-and-shoot (P&S) and the larger sensor full frame (FF). Of course, we will have to enlarge the image from the P&S sensor a factor of six more than the image from the FF sensor to be able to compare them.

11.2 Determinants of Image Appearance

Angle of view. We can maintain the angle of view by scaling the focal length (f) as shown in Figure 11.1. If the lens on the FF camera is set for 60 mm then the P&S must be set to 60 mm/6 = 10 mm.

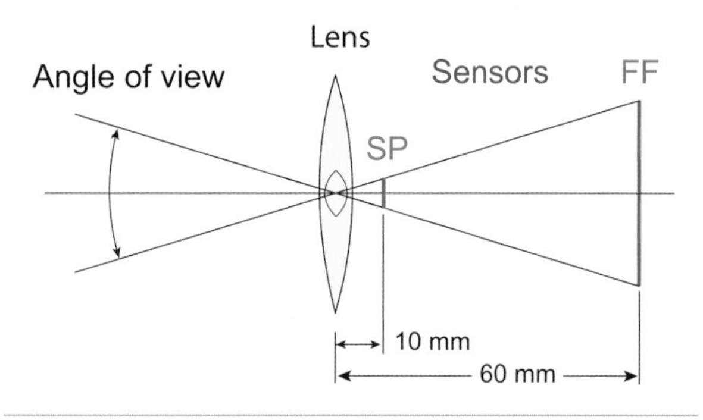

FIGURE 11.1. The horizontal angle of view for FF sensor (focal length = 60 mm) and P&S sensor (focal length = 10 mm).

By scaling the focal length so that it is proportional to the sensor width, we have obtained an image that is similar in one way, but it may not really be equivalent when the other properties are considered. To understand how to make an equivalent image with the P&S and FF cameras, we need to investigate which properties can be scaled and which cannot. Of course, nothing about the real world changes when we use a different sensor, and, in particular, there is no change in

- the desired distance between the object and the lens,
- the size of the object to be photographed,
- the wavelength of light.

With these invariants in mind, we can proceed through the list of image properties.

Perspective. This is an easy item to address because "perspective" depends only on the position of the camera lens and is unaffected by the focal length. In Chapter 4, we showed that the focal length has no effect on the perspective and only determines the size of the image. No matter what size camera/sensor we use, we have only to position the lens in the same place. The idea that we can change perspective by switching to a wide-angle lens is, therefore, incorrect. The perspective only changes when we move relative to the subject.

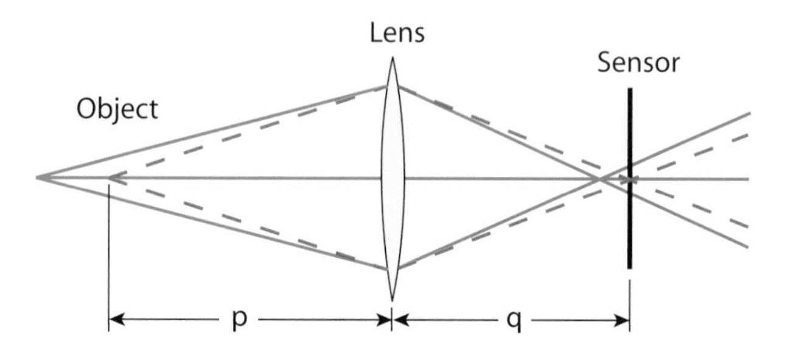

FIGURE 11.2. Light rays from an object to a detector.

Depth of field. Actually, only objects at one distance from the camera are in perfect focus, but the human eye has limitations, and objects at other distances may appear to be in focus in a photograph. The range of distances where things are in acceptable focus defines what we mean by *depth of field* (DoF). Figure 11.2 shows that a point source at a distance p in front of the lens is focused exactly at a distance q behind the lens, i.e., the red dotted lines come together at the sensor. The blue lines from another point, more distant from the lens, are focused at a point in front of the sensor, and when they reach the sensor they have diverged to define a circle. Rays from an object closer to the lens than distance p would focus behind the sensor and would also define a circle on the sensor. These circles are called *circles of confusion*, CoC. If the CoC is small enough, the object is said to be "in focus" and the distance to the object falls within the DoF. The subjective part of this calculation involves deciding on an acceptable CoC. After that is done, the remainder of the calculation is simple, albeit tedious, geometry.

However, there is a caveat. The type of DoF computation depends on the way the photographs will be viewed. In principle, a photograph should always be viewed from its proper perspective point. That is to say, the angle subtended by the photograph at the eye should be the same as the field of view of the lens used to take the photograph. Therefore, photographs taken with a wide-angle lens should be held close to the eye so as to fill much of the field of view, while telephoto photographs should be held farther away. It is sometimes forgotten that perspective in a photograph depends only on the position of the lens relative to the subject (object).

So how will the photographs be viewed? In fact, prints are usually viewed from a distance of 10 in. to 15 in. regardless of the focal length of the lens used to make the photograph. When mounted prints are viewed, observers typically stand about the same distance from all prints. Observers generally don't know what focal-length lens was used, and they simply react to the apparent distortions present when a wide-angle photograph is viewed from a distance greater than the perspective point. Similarly, there is an apparent flattening effect when telephoto photographs are viewed from a distance too close to the eye. Also, in photographic exhibits, the audience often remains seated at the same distance from all prints or from the projection screen. Under usual viewing conditions, it is appropriate to compute the DoF with a constant CoC in the image regardless of the focal length of the lens.

So what is the CoC? This choice depends on how good the human eye is at resolving closely spaced dots or lines. Various authors quote the acuity of the human eye in the range of 0.5 to 1.5 minutes of arc ($1^\circ = 60'$ of arc). To put this information in context, I note that 1 minute of arc corresponds to the diameter of a dime (18 mm) viewed from 68 yards, or 1 part in 3500. Of course, the contrast in photographs is less than that in eye tests that involve black lines on white backgrounds, so we need to relax this requirement a bit. I will follow standard practice and choose 1 part in 1500 as the defining angle. (This is the most conservative value suggested by Canon, although the Zeiss formula uses 1730.) What this means is that the CoC for a given sensor is taken to be the sensor's diagonal length (Diag) divided by 1500. For the FF sensor (or 35 mm film) this gives a CoC of about 30 μm, while for the P&S sensor the CoC is only 5 μm. There is nothing sacred about this choice of CoC. It may be fine for an 8 in. × 10 in. print viewed from 13 in., where the horizontal field of view is only 42°, but for a 13 in. × 19 in. print that will be viewed from the same distance, the divisor probably should be scaled up from 1500. Note that for 8 in. × 10 in. prints, the CoC on the print is only about 0.2 mm. The question of perceived image quality is discussed in detail in Chapter 17.

I begin with the distance to the object (p), the selected focal length (f), and the CoC for each sensor size. This provides enough information for the calculation of the DoF for each value of the lens aperture δ or the F-number (N) since the aperture equals the focal length divided by the N, i.e., $\delta = f/N$. The necessary DoF equations, derived with geometry, are well known and relatively simple, and the results are also well known. Namely, the predicted DoF increases without limit to cover all distances as the N is increased without limit even though the actual resolution will be degraded by diffraction effects. Therefore, the maximum useful DoF is limited by diffraction, and it is necessary that we quantify diffraction effects before completing this calculation.

Diffraction broadening. Diffraction is ubiquitous in nature, but it is usually considered to be a curiosity. Sometimes diffraction is so obvious that it cannot be ignored. We have all seen the rainbow of colors that appear when white light is diffracted from the surface of a CD or DVD. This occurs just because the rows of data pits are separated by distances close to the wavelength of light. (Actually, the separation on a DVD is 650 nm, the wavelength of red light.) When we refer back to first principles, we see that everything having to do with the propagation of light depends on the fact that light is an electromagnetic wave. Diffraction effects become noticeable when light waves interact with small apertures where all of the light rays are constrained to be close together.

The distribution of intensity on the sensor for a point source of light depends on the size and shape of the aperture. For a circular aperture, the image has a bright central dot encircled by faint rings as shown in Figure 5.2. This dot, the Airy disk, is well known in astrophotography. The radius at which the intensity of the central spot reaches zero, i.e., the first dark ring, serves to define the diffraction spot size. With this definition, the diameter of the diffraction spot on the sensor is $2.44\lambda N$, where λ is the wavelength of the incident light and N is the F-number ($f/\#$).

The circle of confusion (CoC) is the key to this analysis. I stated earlier that the CoC has been chosen to define the region where the focus is good enough to be included in the DoF. Now I assert that high resolution in the DoF region also requires that the diffraction spot be less than or equal to the CoC. This upper limit on the diffraction spot size requires that $2.44\lambda N = \text{CoC} = \text{Diag}/1500$, and we immediately have a determination of the maximum useful N for any wavelength λ. Visible light has wavelengths in the range 400 nm (blue) to 700 nm (red); and, therefore, diffraction effects are greater with red light. For the calculation of N, I have selected $\lambda = 555$ nm, the wavelength that corresponds to the region of maximum sensitivity of the human eye. The result is surprisingly simple. To compute the largest useful N value for any camera, simply determine the diagonal dimension of the sensor in millimeters and multiply that number by 0.5. The computed values of N, CoC, and scaled f for various sensor sizes are shown in Table 11.1.

Here f_{FF} refers to the focal length of the lens on the full-frame camera that serves as the reference. The important point is that the value of N computed for a given sensor size is the greatest value that can be used without degrading the image. This calculation applies to an ideal lens where the performance is diffraction-limited. Real lenses have various aberrations and usually show the best resolution at a couple of N stops from the largest aperture. However, as N is increased (aperture is deceased), tests show that performance decreases primarily because of diffraction. If a photograph is cropped, the CoC should be adjusted to the new diagonal, i.e., reduced.

TABLE 11.1.
F-numbers (N) and circles of confusion (CoC) for some common sensors.

Sensor/ film type	Sensor (Daig/mm)	F-number (N)	Coc (μm)	f (Focal length)/f_{FF}
1/2.5"	7.18	3.5	4.8	0.17
1/1.8"	8.93	4.4	5.9	0.21
2/3"	11.0	5.4	7.3	0.25
4/3"	22.5	11	15	0.52
APS-C	28.4	14	19	0.65
35 mm (FF)	43.4	21	29	1.0
6 × 7	92.9	45	61	2.1

Given the shooting distance (p), the focal length (f) and CoC for each sensor size, and the maximum N from the diffraction computation (Table 11.1), we have everything required to determine the DoF. For this calculation I have used a DoF equation that is appropriate for a constant CoC in the image. This equation is derived in standard optics texts, and I have chosen an approximate form that is accurate when the distance p is much greater than the focal length f and the DoF does not extend to infinity:

$$\text{DoF} = \frac{2 \cdot p^2 \cdot N \cdot \text{CoC}}{f^2} = \frac{2 \cdot p^2 \cdot \text{CoC}}{f \cdot \delta}.$$

What we find is that the DoF is the same for all sensor sizes because of scaling of the focal length and F-number. Also, the lens aperture δ remains the same regardless of sensor size. For example, suppose that the image of a 2 ft. object completely fills the diagonal dimension of the sensor when photographed from 5 ft. For a FF camera this requires a focal length of about 100 mm. The optimum F-number is 21, the aperture is 5 mm, and the DoF turns out to be about 10 in. In contrast to this, the P&S camera with the smallest sensor achieves an equivalent image with an 18 mm focal length and an F-number of 3.5.

Shutter speed/exposure. This final property is very interesting. Suppose that the FF camera in the last example required 1/30 s at an ISO sensitivity of 800 for an adequate exposure. The P&S camera with F-number 3.5 admits $(21/3.5)^2$, or 36 times more light and can achieve the same exposure with a shutter speed of 1/1000 s. This calculation depends on the idea that the area of the aperture quadruples when the F-number decreases by a factor of 2. On the other hand, the equivalent photograph requires that we use the same shutter speed to make sure that motional blur is the same. Therefore, in order to obtain the proper exposure, we must decrease the ISO sensitivity of the P&S camera by a factor of 36 to obtain an ISO number of 22. As it happens, most cameras cannot go this low in sensitivity, but maybe we can select 50. This, of course, demonstrates that small sensors do not have to perform well at high ISO settings in order to obtain an equivalent image with maximum useful DoF.

11.3 Comments and Conclusions

I have demonstrated that sensors with different sizes can be used to make identical images if the appropriate ranges of focal lengths, F-numbers, shutter speeds and ISO sensitivities are available. This also assumes that the lenses have sufficient resolving power and the pixel density is high enough not to degrade resolution. The main points are the following:

- The sensor size alone determines the maximum useful F-number (N); and, in fact, the maximum F-number for high-resolution photography is given by 0.5 × (sensor diagonal in mm).
- We can take essentially identical photographs with any sensor size by scaling the focal length, the F-number, and the ISO sensitivity.
- The smallest sensor we considered (denoted 1/2.5 in.) achieves maximum useful DoF at an F-number of $N = 3.5$ while the full-frame 35 mm sensor requires $N = 21$ for a factor of 36 difference in transmitted light. If the full-frame sensor gives the same signal-to-noise ratio) at an ISO sensitivity of 1600 as the small sensor does at ISO 80, the small sensor can still use a higher shutter speed if necessary at maximum DoF. A P&S sensor that could give low noise at ISO 800 or 1600 would appear to have a real advantage over FF sensors.
- If maximizing the DoF is not the aim, larger sensors clearly win because of their ISO-sensitivity advantage. A fast lens ($N = 1.4$) with a full-frame detector is impossible to match with a small sensor. Probably $N = 1$ is the maximum aperture we can expect with a small sensor, and few companies at present even offer $N = 2$. The take-home lesson is that small sensors should be coupled with large-aperture lenses, i.e., small N values. Also, small sensors that support large ISO sensitivities should be sought. The vendors are showing some interest in higher sensitivities, but larger apertures are in conflict with their drive to reduce camera sizes.
- What about pixel count? Doesn't that also influence resolution? Each pixel typically contains one sensor element, so nothing smaller than one pixel can be resolved. It is actually worse than that, because in most sensors each pixel only records the intensity of one color. It is then necessary in hardware or with a RAW converter program to combine information from neighboring pixels to assign three color values (R, G, and B) to each pixel site. As a

rule of thumb, we should double the size of a pixel to approximate its effect on resolution. Therefore, when the number of RGB pixels on the diagonal divided by two approaches 1500, the number we have used to define the CoC, and pixel size severely limits the possible resolution. A 10-megapixel sensor with a 2 : 3 aspect ratio has about 4600 pixels along the diagonal, and pixel size already influences resolution (see Chapter 17).

- The fundamental limitation on DoF is imposed by diffraction effects. Does technology offer any hope for circumventing the diffraction barrier? The answer sounds almost like science fiction: In principle, materials with a negative refractive index (NIM) make "perfect" lenses that are not limited by diffraction (see [Smith et al. 04]). No naturally occurring NIM exists, but artificial materials (metamaterials) have been fabricated that display negative refractive indices for microwaves, which have much longer wavelengths than visible light. In 2005, a NIM material, consisting of minuscule gold rods imbedded in glass, was reported to work with near IR wavelengths. No NIMs are in sight for visible light, and if one is constructed it will probably only be effective at one wavelength—but it is fun to imagine what it could do. NIMs are hot topics in the optics world, and there are many papers exploring their strange properties.

11.4 An Illustration of Maximum DoF with Different Sensors

I have compared two readily available digital cameras with very different sensor sizes (a Canon S80 8-megapixel camera and a Canon 10D 6-megapixel camera) by photographing a table-top scene with each camera. The location of the camera lenses relative to the subject were nearly identical and the focal lengths were adjusted to give the same angle of view. In each case, 6 in. in the plane of focus represented about 40% of the length of the sensor diagonal and the lens-to-object distance was about 32 in. In addition, the F-numbers (N) were computed as described above by multiplying the sensor diagonal by 0.5. The results are shown in Figures 11.3 and 11.4. It was necessary to crop these photos to give the same aspect ratios since the APS sensor is 3 : 2 while the 1/1.8 in. sensor is about 4 : 3. The computed parameters for the S80 were f = 19 mm and N = 4.4; but I settled for f = 20.7 mm, the maximum focal length, and N = 5.3, the minimum value available at full zoom. For the 10D, the computed values were about f = 56 mm and N = 14. The focal length was set to approximately 57 mm, and N = 14 was approximated by the available N = 16 value. The ISO sensitivity was set to 50 on the

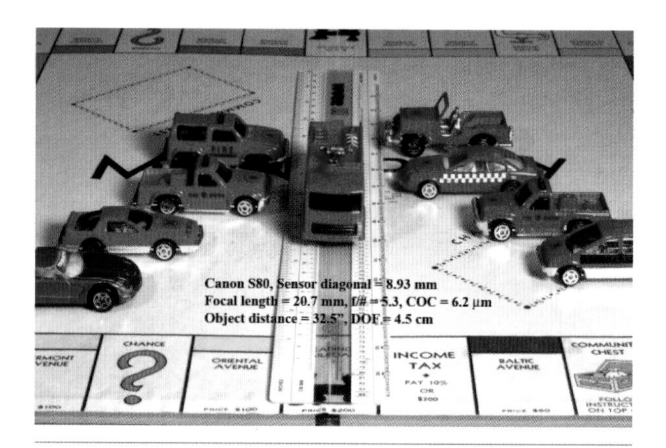

FIGURE 11.3. Photograph taken with a Canon S80 8-megapixel camera. The maximum focal length for the lens was used.

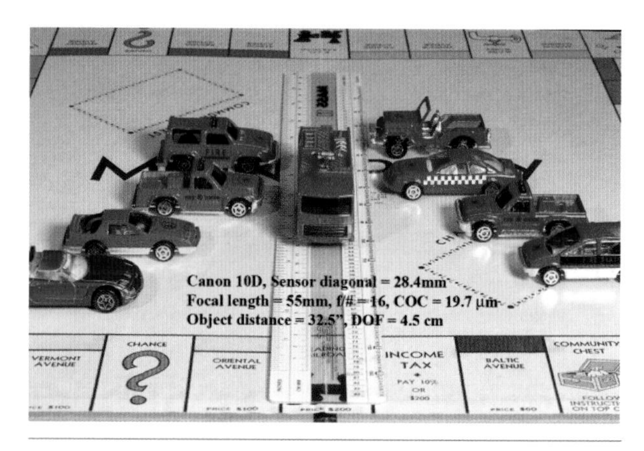

FIGURE 11.4. Photograph taken with a Canon 10D 6-megapixel camera with a 28–135 mm zoom lens. The focal length of the lens was adjusted to match the angle of view of the S80.

S80 and 800 on the 10D to compensate for the difference in N values. Since the ratio of light intensity at the sensors was about $(16/5.3)^2 = 9$, ISO 800 was a higher value-than necessary to make an equivalent image.

The photographs appear to be almost identical when enlarged and examined on a computer screen, and the DoF was found to be roughly equal to the computed value of 4.0 cm. Since the S80 image showed signs of in-camera sharpening, the 10D image was given minimal sharpening to account for the antialiasing filter (Photoshop unsharp mask with radius = 0.3 and strength = 300). The 10D image showed more noise than the S80 image because of the much higher ISO value used. Also, the S80 image exhibited slightly higher resolution, probably because of the larger pixel count.

So what happens when the F-number N is increased? This question was investigated as follows. An image was obtained with the S80 at $N = 8$, the highest available value. Also, an image was obtained with the 10D at $N = 22$. As expected, both of these images are slightly softer in the plane of best focus than the images obtained at the computed N values. Therefore, the CoC values and the computed N values are reasonable.

One obvious conclusion is that the S80 images are remarkably good, but the range of settings is severely limited. At full zoom, the entire F-number range ($N = 5.3$ to 8) suffers from diffraction-limited resolution, and there is no possibility of limiting the DoF if that is desired.

Further Reading

R. Kingslake. *Lenses in Photography: The Practical Guide to Optics for Photographers.* Rochester, NY: Case-Hoyt Corp., 1951.

G. R. Fowles. *Introduction to Modern Optics, Second Edition.* New York: Dover, 1989.

V. Bockaert. "Sensors." Digital Photography Review. Available at http://www.dpreview.com/learn/?/Glossary/Camera_System/sensors_01.htm, 2009.

the lens to the object. This explains the need for long lenses to photograph distant animals and birds. In the absence of a long lens, the photographer has no recourse but to get closer to the subject.

At the opposite extreme, e.g., plants and insects close-up, the magnification is often 0.25 or greater. Strange as it may seem, in this situation, magnification can be increased by decreasing the focal length. This is, in fact, the effect of "close-up" supplemental lenses. The magnification can also be increased by increasing the distance from the lens to the sensor, i.e., adding extensions. These may sound like arbitrary rules, but, in fact, they follow directly from the simple conjugate equation.

Here, again, science unifies apparently disparate sets of observations, and it is worthwhile to spend a few minutes to see how everything fits together in conformity with the laws of optics. This will require a few equations. If equations are not helpful for you, please skip to the next section. For simplicity I will begin with a single thin lens (Figure 12.1), but the results can easily be generalized to compound lenses of any complexity. (Note that the right triangle formed by ends of the object and the center of the lens is similar to the right triangle formed by the ends of the image and the center of the lens. Therefore, (image height)/(object height) = q/p.)

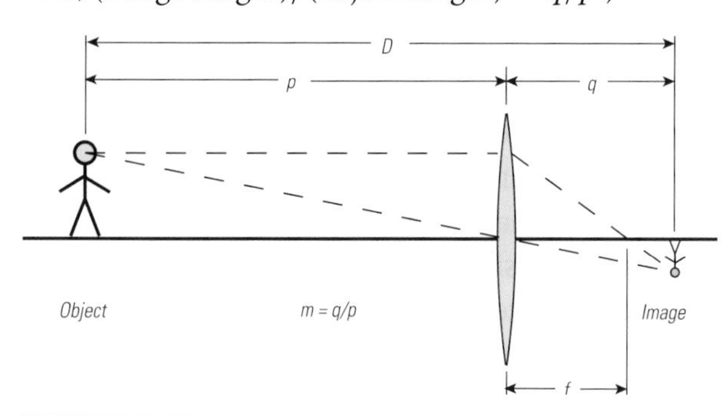

FIGURE 12.1. Magnification with a single thin lens.

As shown in Chapter 7, the working distance, p, and the image distance, q, are related to the focal length, f, by the conjugate equation:

$$\frac{1}{p}+\frac{1}{q}=\frac{1}{f}. \qquad (12.2)$$

In the telescopic limit, the object is located so far to the left that p is much greater than the focal length f. In this limit, light rays from the object form an image in the plane where $q = f$, and the magnification is close to 0. Now, suppose we move the object closer and closer to the lens to increase the magnification. When $p = 2f$, it is easy to show with the help of Equation (12.2) that $p = q$ and $m = 1$. This benchmark arrangement is shown in Figure 12.2.

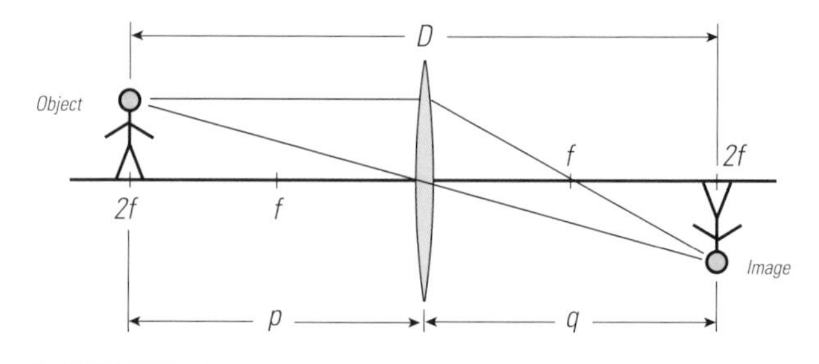

FIGURE 12.2. The arrangement for 1X magnification.

Now, suppose we need a magnification greater than 1X. One way to accomplish this is to move the object closer to the lens, so that $p < 2f$. Of course, the image will not be in focus unless q can be increased to satisfy Equation 12.2. As p approaches the value of f, the rays on the right-hand side become parallel, and the image plane is infinitely far from the lens (on the right). The conclusion is that the entire "macro" range ($m = 1$ to $m = \infty$) is confined between $p = 2f$ and $p = f$. Of course, when p is less than f, the image is virtual, and the lens functions as a magnifying glass. (*Virtual* means that the image appears to be where it does not actually exist, as with reflections in a mirror. A *real* image exists where the rays from a point on an object are brought together in another point.)

This discussion may leave the false impression that focus is always obtained by moving the lens relative to the sensor. This type of focusing, moving the entire lens unit, is known as *unit focusing*. The important parameter in unit focusing is the *extension, or distance added between the lens and the sensor plane*. Focus may also be obtained by the movement of elements inside a compound lens. For example, the front element may be moved relative to the other lens elements, which results in a change in the focal length. Modern compound lenses may, in fact, use both extension and focal length change to achieve focus.

Useful expressions for the magnification can be obtained from Equation (12.2) and the definition $m = q/p$. For example, it is easy to show that

$$m = \frac{q}{f} - 1 \ . \qquad (12.3)$$

Therefore, in the macro range we can increase the magnification by either increasing q (the total extension) or by reducing the focal length f. In the other limit, where p is large and q/f approaches 1, the magnification vanishes.

Another useful expression for m (obtained by multiplying Equation (12.2) by p and rearranging) is

$$m = \frac{f}{(p - f)} \ . \qquad (12.4)$$

In the telescopic limit where p is much greater than f, we find that $m = f / p$ This result justifies our comment that at great distances the magnification is proportional to the focal length.

Equation (12.2) appears to have all the answers, but there is a catch. Practical camera lenses are compound lenses. The object distance p is measured to the first nodal point and the image distance q is measured from the second nodal point as described in Chapter 9. Therefore, the distance from the object to the sensor is usually not equal to the sum of p and q. Furthermore, as mentioned above, the internal movement of lens elements may change the focal length as well as the distance of the lens elements from the sensor. The bottom line is that the simple equations shown above are only useful for compound lenses when the nodal points are close together or when the locations of the nodal points are known.

Another factor that must be considered is the increase in the effective F-number, N_{eff}, relative to the value of N, reported by the camera, that usually occurs when the magnification is increased. This, of course, can greatly increase exposure times and exacerbate the effects of diffraction. It turns out that N_{eff} is related to the magnification m by the equation $N_{eff} = N(1 + m)$ except when magnification results from the use of a supplemental close-up lens. However, there is a caveat. Lenses are characterized by an entrance pupil and an exit pupil. For a symmetrical lens the entrance and exit pupils are equal, but in general they can be quite different. Typically, the entrance pupil is larger than the exit pupil for telephoto lenses, while for wide-angle lenses the exit pupil is larger than the entrance pupil. This can be verified simply by looking into a lens and estimating the apparent size of the aperture stop from the front and the rear. The pupil ratio is defined as

$$p_r = \frac{\text{exit pupil}}{\text{entrance pupil}} \ ,$$

and a more accurate equation for the effective F-number is

$$N_{eff} = N \cdot \left(1 + \frac{m}{p_r}\right) . \qquad (12.5)$$

So where does this leave us? The simple conjugate equation is not very useful for macro photography, the positions of the nodal points for the lens are in general not available, and the entrance and exit pupil diameters can only be estimated. (Actually, these numbers have been reported for a few prime lenses.) The situation is not ideal, but with a modern DSLR we can measure all the quantities needed for macro photography.

Consider the standard set of close-up accessories:

- Macro lens or close-focusing tele-zoom lens
- Supplemental lenses (1–50 diopter)
- Extension tubes or bellows
- Tele-converter (1.4X or 2X)

With a prime lens and any combination of supplemental lenses and extensions in place, the following quantities can be measured (or estimated) at close focus and far focus (∞):

- Working distance, w (distance from object to front of lens combination)

- Distance from object to sensor plane, D
- Magnification, m (from measurement of image height on sensor)
- The effective F-number: N_{eff} (from the automatic aperture setting at fixed shutter speed), and
- Resolution as a function of N_{eff}

The attachments and relevant distances are shown in Figure 12.3.

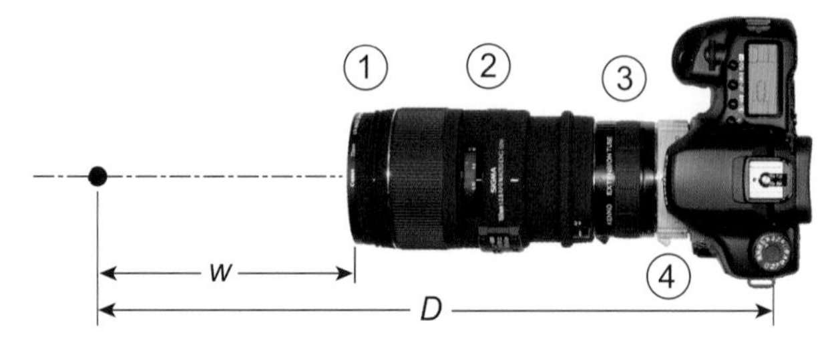

FIGURE 12.3. Illustration of the working distance w and the distance D to the sensor for a camera with accessories: (1) supplementary close-up lens, (2) macro lens, (3) extension tubes, and (4) tele-extender.

The next few sections cover the various combinations and arrangements for increasing magnification and point out their advantages and limitations.

12.2 Supplemental Lenses

Probably the easiest and least-expensive way to obtain close-up or even macro capability with a standard lens is to add a supplementary lens. The simplest of these is a one- or two-element screw-in lens that attaches just like a filter, although, of course, they are *not* filters. These close-up lenses are sometimes called *diopters* or *plus diopters,* because their strength is rated in diopters. (Recall that the optical strength is proportional to the inverse focal length, or $d = 1/f$ and the dimensionless strength in diopters is obtained when the focal length is measured in meters.) If the focal length is measured in millimeters, as is usually the case, the optical strength is $d = 1000\text{ mm}/f = 1000/(f/\text{mm})$ diopters. For example, a 100 mm lens has a strength of $d = 1000/100 = 10$ diopters.

The magnification expected from a camera lens (prime lens) with an attached diopter can be estimated with the simple-lens equations. The basic idea is that the strength of the combination of two closely spaced lenses is just the sum of their strengths in diopters. This result can be expressed either in terms of diopters, as $d = d_{\text{Prime}} + d_{\text{Diopter}}$, or focal lengths:

$$\frac{1}{f_{\text{Combo}}} = \frac{1}{f_{\text{Prime}}} + \frac{1}{f_{\text{Diopter}}} \tag{12.6}$$

Suppose that the prime lens is described by $f_{\text{Prime}} = 100$ mm and a +2 diopter close-up lens is added. The effect is to decrease the combined focal length and hence to increase the optical strength. In this case, the strength becomes 10 diopters +2 diopters = 12 diopters, and the combined focal length is 83 mm. The magnification can easily be estimated when the prime lens is set at the far-focus (∞) position so that $q = f_{\text{Prime}}$. This, of course, means that all parallel rays entering the prime lens will be brought to focus in the plane a distance f_{Prime} from the lens. Also, we know that rays from a point on an object at a distance $p = f_{\text{Diopter}}$ from the close-up lens will be parallel when they exit the plus diopter. As shown in Figure 12.4, the combination of a close-up lens with a prime lens always gives a working distance (w) equal to the focal length f_{Diopter} and a magnification is

$$m = \frac{f_{\text{Prime}}}{f_{\text{Diopter}}} \tag{12.7}$$

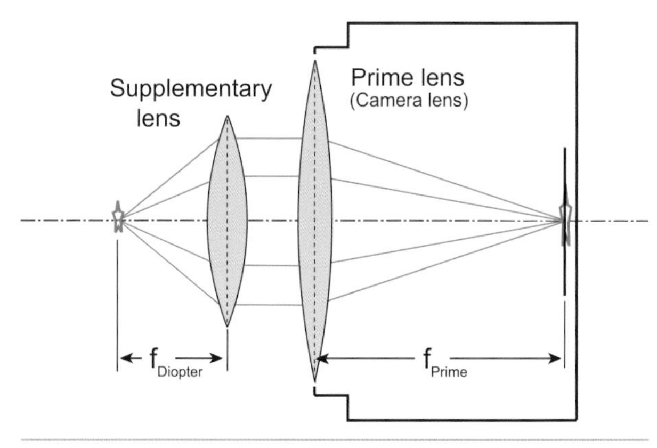

FIGURE 12.4. The operation of a close-up lens when the prime lens is focused at infinity.

Equation (12.7) is useful for standard lenses with limited focusing ranges. For the 100 mm lens with a +2 diopter supplementary lens, we obtain $m = 0.2$. As a test, we attached a +2 diopter lens to a zoom lens and set the focal length to 100 mm. I found that at far focus the working distance was approximately 50 cm and the magnification was $m = 0.20$. When the lens was focused as closely as possible, the working distance was reduced to 16 cm, and I measured $m = 0.27$. As a more extreme example, I attached the +2 diopter to a 105 mm macro lens that is capable of 1X without attachments. Here again, at far focus the working distance was 50 cm and m = 0.2 but at close focus the working distance is only 8.5 cm and $m = 1.24$. With the macro lens it would be crazy not to use the close-focus capabilities, and therefore Equation (12.7) is not very helpful. Still, the close-up lens increases the maximum magnification a bit by reducing the effective focal length. This can be useful when the amount of extension is limited.

Supplementary lenses have important advantages. They are small and easy to use, they have little effect on the effective F-number[1], and their contribution to the magnification increases with the focal length of the prime lens. The main drawback is their effect on optical quality. Inexpensive sets of diopters typically contain single element lenses and it is necessary to stop down the prime lens to maintain acceptable resolution. I have obtained good results with a set of HOYA coated diopters (+1, +2, and +4). In general, however, photographers are well advised to use two-element (achromatic) close-up lenses whenever possible. These are now available from a number of manufacturers. I know from experience that high-strength, single-element diopters, e.g., $d = 10$, severely limit resolution.

I have saved the best for last. It is possible to obtain high-quality supplementary lenses with high optical strengths up to at least +20 diopters. Furthermore, these lenses are widely available at reasonable prices; and many people may already own one or more of them. I am, of course, referring to fairly short-focal-length, large-aperture camera lenses that can be reversed and used as supplementary lenses. A 50 mm, *f*/1.8, Canon lens makes a fine +20 diopter close-up lens and can be easily mounted on a variety of prime lenses by means of "macro coupler" reversing rings. I have also used light-weight Olympus OM system 50 mm, *f*/1.8, and 100 mm, *f*/2.8, lenses as excellent +20 and +10 diopter supplementary lenses. Lenses such as these are available on the used-lens market at very reasonable prices. The combination of a telephoto lens with a reversed 50 mm lens is shown in Figure 12.5.

FIGURE 12.5. The combination of an Olympus 50 mm lens, a macro coupler, and a Canon 70–300 mm lens that permits magnifications up to 6x.

12.3 Extension

Another common method for making any lens focus "closer" is to add an extension between the lens and the sensor plane. A convenient way to accomplish this for DSLRs is to use a set of extension tubes. For example, the set sold under the name Kenko includes 12 mm, 20 mm and 36 mm tubes that can be used separately or stacked together to give 68 mm of extension. These tubes contain no glass elements but do provide electrical connections between the prime lens and the camera body for automatic focusing and exposure control. It is also possible to obtain extension bellows with a focusing rail for some cameras. These systems with (automatic) electrical connections are very expensive and are rarely used by amateur photographers.

So, how much good does added extension do? As an example, consider the Canon 50 mm $f/1.8$ II lens. At close focus, the distance from the object to the sensor plane (D) is approximately 44 cm, the working distance (w) is 34 cm, and the measured magnification (m) is only 0.15. Close focusing with this lens only adds a small amount of extension, and q remains approximately equal to the focal length. Now I add an additional 68 mm of extension with the entire set of stacked tubes. With this addition, the measured magnifications for close- and far-focus become 1.46 and 1.31, respectively. This is true macro capability, but the combination is not very flexible: and the working distance range is only from 4.8 to 5.3 cm. Furthermore, the effective F-numbers are approximately 2.5 times greater than the values of N reported by the camera.

The following, often quoted, equation relates the magnification to the added extension: m = (added extension)/(focal length). This equation is, of course, just Equation 12.3 with $q = f + \textit{added extension}$. It gives some guidance for standard prime lenses, but in real-world macro photography, where one usually works with zoom lenses and/or macro lenses having considerable built-in extension, the equation is almost useless. The Sigma 105 mm macro lens proves the point. With 68 mm of added extension, the magnification ranges from 0.66 to 1.89, going from far to close focus because of the considerable extension in the lens mount. The only practical way to characterize a combination of lens and extension tubes is to measure its performance. The macro lens with extension tubes may be practical when magnifications greater than 1X are required, but here again there is considerable loss of light. I estimate the effective F-numbers are 2 to 4 times the reported values.

In general, extension becomes cumbersome as the focal length increases. However, the increase in magnification with an extension tube may be more than expected because the prime lens is likely to achieve close focus by decreasing the focal length. Also, a small amount of extension can be useful with short-focal-length (wide-angle) lenses to obtain striking images of extended objects such as almost-flat flowers. Of course, fisheye and ultra-wide-angle lenses often focus very close without any extension. For example, the photograph in Figure 12.6 was taken with a 15 mm fisheye lens (APS-size sensor).

FIGURE 12.6. A rose photographed in the late afternoon without flash (f/11, 1/30 s); background blur was increased by computer processing.

The conclusion is that extension tubes can be useful and are well worth owning. They are relatively inexpensive and offer increased magnification without decreasing optical quality. Also, they can be a lifesaver when your favorite telephoto lens does not focus close enough. In this situation an added extension tube may well do the trick.

12.4 Teleconverters (Tele- Extenders) for Cameras with Interchangeable Lenses

Teleconverters are lens accessories that are usually mounted between a lens and a camera body in order to increase the effective focal length of the lens. Teleconverters, sometimes called tele-extenders, are negative lens elements (or groups) that are not capable of forming an image on their own. Peter Barlow (1776–1862) recognized that the introduction of a diverging lens element would multiply the focal length while maintaining approximately the same overall length of the lens. The basic idea is the same as that encountered in the design of telephoto lenses, and was even noted by Galileo in about 1609. As shown in Figure 12.7, the focal length of a simple or compound prime lens can be stretched by introducing a negative (diverging) lens after it. The combination produces a simple telephoto lens with the focal length f, and the telephoto ratio is defined as L/f.

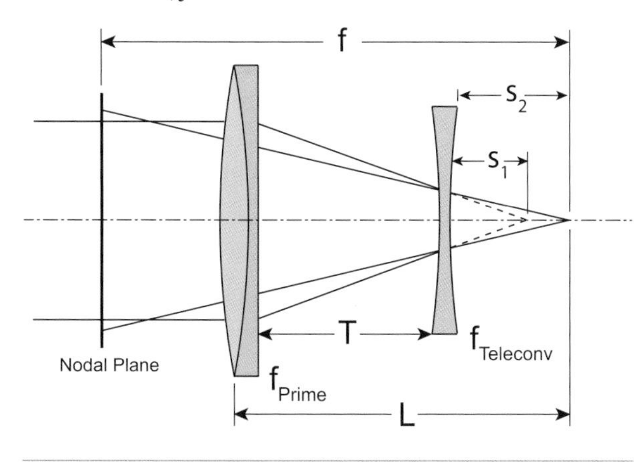

FIGURE 12.7. The effect of a teleconverter or Barlow lens on rays from a prime lens.

When simple lenses with paraxial rays are involved (Chapter 7), the focal length of the closely spaced combination is related to focal lengths of the separate components by Equation (7.6).

In the real world, with long telephoto lenses and rather thick teleconverters, simple equations are not very useful. The important point is that the teleconverter provides a magnification increase by a factor of $m = s_2/s_1$. This means that a 100 mm lens acts like a 200 mm lens at least at far focus, but at any focal distance the magnification is increased by the same factor. Furthermore, this factor increases the magnification already achieved by macro lenses or extension tubes. Therefore, converters act as multipliers of magnification. Teleconverters also preserve the working distance as well as the full focusing range of the prime lens out to infinity. So what is the downside? Everything in photography involves compromise. With teleconverters, we lose light. Since the focal length increases while the aperture remains the same, the F-number must increase. The 1.4X teleconverter costs one stop and the 2X unit costs two full stops. In addition to this, the introduction of additional glass always reduces the quality of the image. The decrease in quality may be small at 1.4X, especially if the lens is stopped down, but with 2X and higher teleconverters the reduction is often noticeable.

Even with some reduction in image quality, the use of a 2X teleconverter may be worthwhile. The Juza photo site has reported excellent macro photography with a 180 mm macro lens combined with a 2X teleconverter (see [Juza 09]). This combination provided increased working distance with impressive resolution at small apertures when stopped down to $f/11$. Of course, diffraction is the limiting factor. The effective F-number is at least twice as great with the 2X teleconverter, and the rule of thumb is that diffraction becomes significant when the F-number exceeds one half of the sensor diagonal expressed in mm, i.e., about $f/14$ for an APS (1.6 crop factor)-type sensor.

12.5 Afocal Teleconverters for Fixed-Lens Cameras

Small cameras and less-expensive cameras usually do not have interchangeable lenses. Therefore, telephoto and wide-angle attachments for these cameras must be attached in front of the camera lens. This is analogous to using a fixed-lens camera to take photos through the eyepiece of a telescope, thus replacing the human eye with a camera. For this to work with any camera lens, an afocal arrangement (shown in Figure 12.8) is required. The magnification in this arrangement is given by $m = f_1 / f_2 = f_{eff} / f_{camera}$ when the camera is focused on infinity. For closer distances the actual magnification measured from the image size on the sensor depends somewhat on the focal length of the camera lens. The same principles can be used to make a front-mounted wide-angle converter.

A focal teleconversion lenses can be quite satisfactory for point-and-shoot cameras and camcorders. They also work well on lenses designed for DSLRs, but tend to be bulky, and they offer no

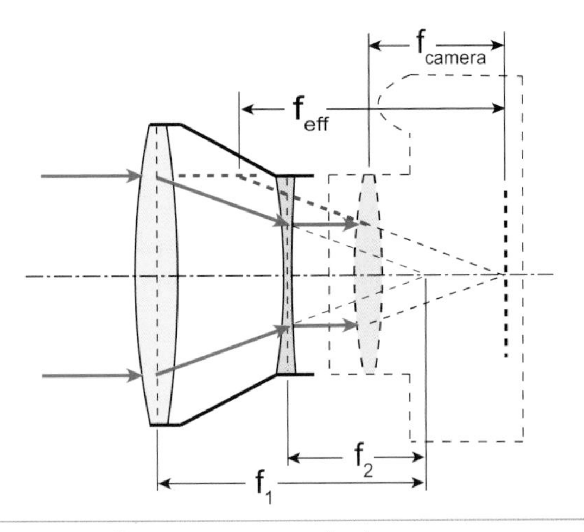

FIGURE 12.8. Front-mounted teleconverter.

advantage over the rear-mounted teleconverters. In fact, there is a serious problem for close-up photography. In contrast to rear-mounted teleconverters, the front-mounted units greatly increase the minimum focusing distance. So there might be an advantage when photographing a bird at 10 ft., but at close distances the maximum magnification is actually decreased. In addition, afocal teleconverters usually lack threads for mounting polarizers and filters, and they may put unacceptable stress on lens barrels. In particular, they are not recommended for lenses with rotating front elements.

12.6 Combinations of Lenses and Extenders for Real-World Macrophotography

I have shown that magnification can be increased by changing the focal length and/or changing the amount of extension. Your best course of action, of course, depends on your application and the equipment at hand. There are pros and cons for the various options. Here is a summary that repeats some previous advice and adds some information about useful combinations of techniques. I am assuming that extreme magnification is not required and that a maximum of 1X suffices with occasional excursions to 3X or 4X.

Macro lenses. It is convenient to shoot everything from birds to small insects with a macro lens on an interchangeable-lens camera. Large apertures (*f*/2.8) and long focal lengths (100 mm, 180 mm, etc.) that permit good working distances and reasonable exposures when there is good light are readily available. The working distance can be increased when needed by adding a 1.4X or 2X teleconverter lens at the cost of reduced light transmission and perhaps image quality. When higher shutter speeds are necessary, supplementary light sources may be required (see Section 12.8). It should be noted that Canon also offers a special-purpose manual-focus macro lens (Photo MP-E 65 mm *f*/2.8) for magnifications from 1X to 5X. Experienced photographers can obtain remarkable results with this lens, but it is not for everyone. I recommend experimentation with magnifications less than 5X before considering the MP-E. Figure 12.9 shows photographs taken with a 105 mm macro lens in good light.

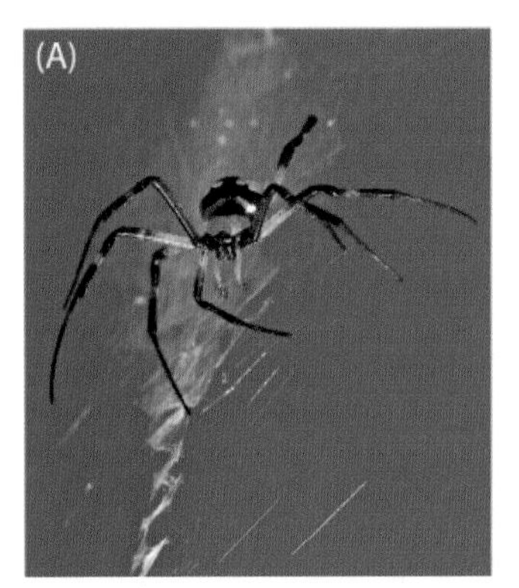

FIGURE 12.9. Photographs taken with a 105 mm macro lens: (a) an argiope illuminated by a bright sky and reflector with a gray card in the background (f/6.7, 1/250 s); (b) a praying mantis illuminated by flash (f/11, 1/15 s).

Supplementary lenses. A diopter increases magnification by permitting closer focusing without much change in extension. The diopter makes up for limited extension by decreasing the effective focal length of the camera lens. Because the increase in magnification is offset by the decrease in focal length, the effective F-number is unchanged and there is little loss of light. On the other hand, the cons are

- a loss of image quality because of the added glass,

- a loss of working distance,
- a limited focusing range.

The image-quality problem suggests it is not a good idea to stack many supplementary lenses together or to use supplementary lenses in combination with rear-mounted teleconverters. However, a single diopter of medium strength, e.g., +2, in combination with a zoom lens can be quite useful when a light mobile setup is needed for butterfly photography.

Extension tubes and teleconverters. Given enough extension, one can obtain any amount of magnification, but this is often not very practical. Large extensions are bulky and very sensitive to vibration. Also, the effective F-numbers are large and long exposures are required. It is usually a better idea to use a moderate amount of extension, i.e., less than the focal length of the lens, to enhance the magnification already obtained with a macro lens or to reduce the minimum focusing distance of a telephoto lens. It is also possible to obtain high-quality macro images by combining the use of supplementary lenses and extension tubes. The effect of the added diopter is to amplify the effect of the extension.

Another popular technique is to use both extension tubes and a teleconverter. The first question is about the best arrangement of the components. We know that a rear-mounted teleconverter increases the effective focal length, but it also multiplies whatever magnification has been obtained by the combination in front of it. The added extension produces magnification equal to the added extension divided by the focal length of the lens at least when the prime lens is focused at infinity. For example, a 105 mm macro lens at far focus has the magnification $m = 0$, but with a 68 mm extension, the magnification becomes $m = 68/105 = 0.65$ (measured value 0.66). When a 1.4X teleconverter is added without any extension, the magnification remains zero at far focus. Suppose we use both a multiplier and an extension. First attach the teleconverter to the lens and then the extension. The result is to increase the focal length to 1.4×105 mm = 147 mm followed by the effect of extension, and we obtain a magnification of $m = 68/147 = 0.46$ (measured value 0.47). Now let us reverse the order and first attach the extension, to obtain 0.64. The teleconverter is then attached, and the result is $m = 1.4 \times 0.64 = 0.90$ (measured value 0.93). It is obvious that, at far focus, the best plan is first to mount the extension followed by multiplication with a teleconverter. These calculations are easy and accurate but may be useless because no one does macrophotography with their macro lens focused at infinity.

So what is the real story about the stacking order? The conclusions reached above have some validity for normal lenses with limited focusing ranges, but for macro lenses and lenses with a macro mode, the stacking order is far less important. This is to be expected, because macro lenses have a lot of extension built in at close focus. Consider again the 105 mm macro lens that gives 1X magnification at close focus. When I mounted 68 mm extension on the lens, followed by 1.4X, I obtained $m = 2.82$ at a working distance of $w = 7$ cm. The reverse, 1.4X followed by 68 mm, gave $m = 2.71$ and $w = 9.5$ cm. It makes sense always to attach the teleconverter last (closest to the camera) even though the order might not make much difference at close focus. An extensive set of measurements for the Sigma 105 mm macro lens with various combinations of extension tubes, diopters, and a teleconverter is presented in Appendix E. An immediate conclusion is that an extension tube in combination with a teleconverter give a versatile, high-magnification setup.

12.7 Special Problems Resulting From High Magnification

The accessories and their combinations described in Section 12.6 all permit large magnifications to be obtained. Magnification is, of course, required for macrophotography, but that is only the beginning. Most of the arrangements require that we work with large F-numbers. Additional requirements may include

- photographing moving objects that require high shutter speeds,

- optimizing F-number for depth of field and resolution at high magnifications,
- enhancing the depth of field within the restrictions imposed by diffraction.

A number of decisions are required here, and the restrictions become severe at the highest magnifications. As always, photography involves compromises. Section 12.8 provides some suggestions for dealing with the most common problems.

12.8 Lighting for Close-Up Nature Photography with High Shutter Speeds

For butterflies and other large insects, the necessary magnification is not very high, and in bright sunlight there may not be a problem. I recommend the following:

- a long focal-length macro lens,
- a zoom lens with a diopter, or
- a close-focusing telephoto lens with a short extension tube if necessary to reduce the minimum focusing distance.

This can be very effective for butterflies and small birds, as shown in Figure 12.10. If the subjects are much smaller, e.g., ants and aphids, the magnification and the effective F-numbers may become large, so that high shutter speeds will usually require supplemental light sources.

FIGURE 12.10. Tropical butterfly photographed with a 70–300 mm (at 240 mm) hand-held IS lens (f/6.7, 1/90 s); supplemental flash enhanced the illumination.

FIGURE 12.11. Aphids and intruder taken with a 105 mm macro lens with an extension (f/11, 1/125 s).

The aphids in Figure 12.11 were illuminated with bright sunlight and a flash. The only way to stay mobile is to use flash attachments mounted on the camera lens or on brackets. A number of vendors sell macro ringlight flash units that mount on the camera lens. These units are good for clinical work, photographing coins, etc., but they give flat, shadowless lighting that is not natural. The best bet for portable macro lighting in the garden is to use brackets for mounting one or perhaps two flash units close to the lens. Kirk, Manfrotto, and others offer a variety of brackets.

A Manofrotto telescoping camera/flash bracket is shown in Figure 12.12. Another possibility is to

FIGURE 12.12. Flash with off-camera extension cord and a mounting bracket.

mount slave flash units near a target area so that they can be triggered either by a camera-mounted transmitter or the light from a flash unit. Several inexpensive slave units can be used to illuminate a subject as well as the background. Since some flash units emit preflashes for focusing or red-eye reduction, the slave units must be able to select the appropriate flash for the exposure.

12.9 The Optimum F-Number for Macrophotography

First recall that the depth of field becomes very shallow at high magnifications. For close-up and macrophotography the depth of field can be expressed as

$$\mathrm{DoF} = 2 \cdot \mathrm{CoC} \cdot N_{\mathrm{eff}}/m^2. \tag{12.8}$$

Here CoC is again the circle of confusion, N_{eff} is the effective F-number, and m is the magnification. The calculation of the depth of field proceeds by selecting an acceptable value of CoC to define resolution that is "good enough." The choice of CoC is subjective, but I previously suggested that a reasonable value would be the sensor diagonal (Diag) divided by 1500. This is probably greater than necessary for small prints and too small when large prints are planned. In general we denote the divisor by Div. The maximum resolution possible and the minimum depth of field would require that the CoC be set equal to approximately twice the pixel width for the sensor in use. For example, an APS-C-type sensor with 10.1 megapixels has Diag = 28 mm and a pixelpitch of 5.7 μm. The rule Diag/1500 gives CoC = 19 μm, while twice the pixel dimension is 11 μm.

Next, a value of N_{eff} is chosen to obtain the desired DoF. As discussed in Chapter 11, however, the diameter of a diffraction-broadened spot in the image is approximately $2.44 \cdot \lambda \cdot N_{\mathrm{eff}}$, and this quantity controls the resolution when N_{eff} is large. If the maximum DoF is desired, then the maximum N_{eff} value consistent with the CoC must be determined. This is easily accomplished by setting the diameter of the diffraction spot equal to the CoC:

$$2.44 \cdot \lambda \cdot N_{\mathrm{eff}} = \mathrm{CoC} = \mathrm{Diag/Div}. \tag{12.9}$$

The value of N to set in the camera is obtained by combining Equations (12.5) and (12.9):

$$N = \frac{\mathrm{Diag/Div}}{2.44\,\lambda\,(1 + m\,/\,p_r)}\,. \tag{12.10}$$

The substitutions, Div = 1500, λ = 555 nm, and p_r = 1, produce a useful expression for N:

$$N = \frac{\mathrm{Diag}}{2 \cdot (1 + m)}\,. \tag{12.11}$$

Equation (12.11) is similar to the rule of thumb presented in Chapter 11 except for the inclusion of the magnification.

As an example, consider a setup with Diag = 28 mm and m = 1. With the resolution specified by Div = 1500, we find that the maximum allowed value of N is 7, i.e., *f*/7. The corresponding DoF is about 0.5 mm. This may seem restrictive, but consider what happens as the magnification increases. When m = 2, the maximum N value is 4.7 and DoF = 0.13 mm; when m = 4, the maximum N is 2.8 and DoF = 0.03 mm. If one is willing to sacrifice resolution for DoF then a smaller value of Div can be chosen, but the depth of field at large magnifications will still be paper-thin. For this reason, it is very hard to take acceptable photographs with a magnification above 4X. Patience, good technique, and a lot of practice are required. The loss of resolution with increasing F-number is demonstrated in Figure 12.13. The *f*/4.5 image has the smallest DoF while the *f*/22 image is beginning to show the effects of diffraction broadening.

FIGURE 12.13. A Sundew leaf at f/4.5, f/8.0, and f/22 taken with a 150 mm macro lens and flash illumination.

12.10 Expanding the Depth of Field

So how can the DoF be expanded? The laws of physics cannot be violated, and there is no way to capture an image that simultaneously offers high magnification (>1X) and large DoF. There are, however, ways to combine a series of photographs obtained with different focal points to produce an image that appears to have a much larger depth of field than any one of the component images. This is a much more difficult task than stitching together images to increase the field of view or even combining images obtained with different exposures to generate an image with a higher dynamic range (HDR). When the focal point is shifted, objects that are out of focus may shift, become larger, and become more fuzzy. Combining a set of such images requires judicious cropping and pasting (or erasing), and the attempt may fail to produce an acceptable image.

Fortunately for photographers, very clever programmers have devoted a lot of time to automating the job of compositing images with differing focus points. The programs HeliconFocus and CombineZM (freeware) are results of this work. These programs have been reviewed and compared online, and tutorials are available for the freeware program. A photographer must produce a set of images with different points of focus of focus and store them in a "stack." The stack is then specified as input for the compositing program. The program rescales the images as needed and aligns them. The sharp parts of each image are selected and then blended to make the final composite. An example is shown in Figure 12.14.

The production of a satisfactory composite at high magnification may require 30 or 40 images and, depending on the amount of overlap of image components at different distances from the lens, the automated result may be unsatisfactory. Sometimes manual manipulation of the composite images or selection of the images to remove from the stack is necessary. This requires experience and trial and error. It is, of course, essential to capture enough images

FIGURE 12.14. A fly trapped on a Sundew leaf; the left-hand shot was taken at f/5.6, 1/8 sec; the image on the right was constructed by CombineZM software from eight photographs.

that the regions of good focus overlap. I have found that a focusing rail with micrometer adjustment is much more satisfactory for stepping through the focusing range than trying to change the focus of the lens. An arrangement for macrophotography with a Manfrotto focusing rail is shown in Figure 12.15.

12.11 Conclusions

Photography at its best gives us new ways to see the world. Macro photography permits us to see things that are almost invisible to the naked eye, or at least things that escape attention. Without high magnification I would not know the beauty of a single deadly drop on a Sundew leaf (shown in Figure 12.14) or be able to see the interaction of aphids with other insects. Also, the details of butterflies and spiders would elude me,

FIGURE 12.15. A focusing rail that is convenient for capturing images with different focal points.

and the tiny flowers of the Sundew plant would remain unappreciated. All of these examples are from nature, and nature seems to provide the best hunting ground for the macro photographer. Through natural selection, the biosphere creates a vast array of structures, ranging from nanoparticles to whales, by construction with one atom at a time. And in the micro and macro (hundreds of microns) range, nature is vastly more prolific and interesting than the creations of technology. So for the time being there is not much interest in the macro photography of inanimate object outside the science laboratory. Thus, coins and paper money only serve to demonstrate magnification and to test resolution.

Whether or not the macro photographer feels limited to the world of nature, macro photography is now accessible to almost anyone with enough interest. The least expensive digital cameras offer macro modes that are surprisingly good though with very short working distances. For those able to afford more expensive equipment, the capabilities are truly impressive. Some DSLRs provide "live view" that permits the live image on the sensor to be magnified ten-fold to aid in achieving focus. The image can also be displayed in real time on a computer screen, and this opens the possibility of video at the micro scale. Macro lens are getting better and better, and computer-processing methods such as image combination open new possibilities every year. I think it is a great time to investigate this area of photography.

Further Reading

J. Shaw. *John Shaw's Closeups in Nature.* New York: AMPHOTO, 1987.

Canon USA, Inc. *Close-Up & Macro Photography.* Canon Workshop Series. Lake Success, N.Y.: Canon USA, Inc., Camera Division, 1996, Publ. Code No. 9-01700.

M. A. Covington. *Astrophotography for the Amateur, Second Edition.* Cambridge, UK: Cambridge University Press, 1999. (Deals with afocal coupling of lenses to telescopes.)

F. Frankel. *Envisioning Science.* Cambridge, MA: MIT Press, 2002. (Deals with photography of small inanimate objects.)

CHAPTER 13

Do We Need Filters Anymore?

Filters allow added control for the photographer of the images being produced. Sometimes they are used to make only subtle changes to images; other times the image would simply not be possible without them.

—WIKIPEDIA

Digital cameras are excused from most of the color conversion filters, since you dial these in as white balance settings.

—HTTP://WWW.KENROCKWELL.COM/TECH/FILTERS.HTM

13.1 Introduction

So you have a digital camera and it takes great photographs. Could you do any better by using filters? As the lawyers say, it depends. There are two types of filters you might consider. First and most important is the polarizing filter. This can give control over reflections from glass, water, other shiny surfaces, and light scattered from the sky. Then there are neutral-density (ND) filters. These special purpose filters are much less important but can help you deal with scenes where there are extreme variations in brightness, such as a dark foreground with a bright sky. They can also help you emphasize motion by permitting long exposures. The remaining type of filter is the color filter. The effects of these filters can be simulated by digital processing and are no longer necessary except for UV and IR photography where visible light must be blocked. This chapter explains in detail how ND, color, and polarizing filters work. I will start with definitions and will go rather far into the theory. There is also a bit of history you might find interesting.[1]

Photographic filters are transparent plates that are designed to modify the properties of light before it reaches the film or sensor of a camera. Filters are usually mounted in front of the camera lens,

FIGURE 13.5. Combined HDR image (left) and single image (right).

The example in Figure 13.5 shows a dinning room with very bright windows on a late summer morning. The image on the left combines three photographs obtained with the exposure compensations of −2, 0, and +2, while the image on the right shows a single photograph with an exposure compensation of +2. The blending of images was accomplished with Photomatix.

Color filters. As noted, color filters have been very important in photography for a century but have been largely replaced by white-balance adjustments in digital photography and simulated filters in the conversion of color images to monochrome. Figure 13.6 shows what color filters do.

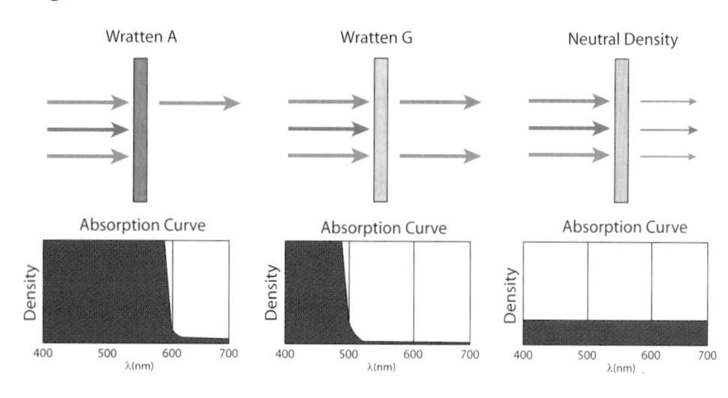

FIGURE 13.6. Color transmission by absorption filters: red (A), yellow (G), and ND.

The filters are, of course, named for the color of light that is transmitted in addition to detailed numbering in the Wratten system. Figure 13.7 shows the conversion of a color image to monochrome with enhanced transmission of yellow and red light by means of the "Channel Mixer" feature of Photoshop software. Of course, subtle tonal changes in color images can be obtained through the use of "warming" filters, "sunset filters," etc. but, in principle, any such tonal changes can be simulated through software manipulations.

13.3 Polarization Filters

Polarization selection continues to be extremely important in photography, and its effects cannot be easily simulated in software. Understanding this phenomenon requires a consideration of the interaction of light with matter. For our purposes, light can be considered to be an electromagnetic wave with electric (E) and magnetic fields (H) perpendicular to the direction of propagation. The direction of the electric field defines the direction of polarization. The ray shown in Figure 13.8 is plane-polarized (transverse magnetic, or TM), has the wavelength λ, and is propagating to the right with speed *c*. The basic idea is that molecules tend to absorb light more strongly when their long axis is parallel to the direction of polarization. The effect is pronounced for molecules aligned in crystals, and the orientation-dependent absorption is known as dichroism. In general, some kind of alignment of atoms or molecules is required for dichroism. Efficient and inexpensive linear polarizers are now available

Kodachrome slide

Simulated yellow filter

Simulated red filter

FIGURE 13.7. The conversion of a Kodachrome slide to monochrome with simulated yellow and red filters.

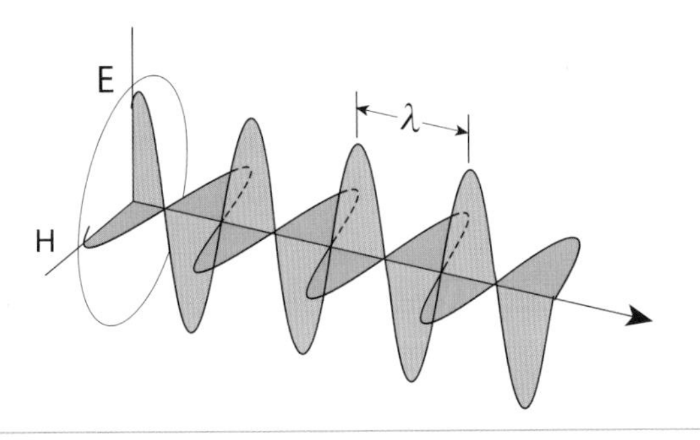

FIGURE 13.8. Polarized light with the wavelength λ.

because of the insight of E. H. Land, the founder of the Polaroid Corporation.[3]

Unpolarized light consists of a collection of rays like that shown in Figure 13.8, but with random directions for their electric fields. In the following figures, I show only the electric-field amplitudes since electric fields dominate absorption processes. Notice that, in Figure 13.9, the electric fields initially lie in planes distributed around the *z*-axis. When the bundle of rays encounters the polarizer, the *y*-components of the electric fields are absorbed, and the *x*-components are transmitted. The result is a filtering process that passes an *x*-polarized beam. Figure 13.9 implies that the absorbing units are aligned in the *y*-direction. A perfect polarizer would pass 50% of the intensity, but, of course, there are reflections and other losses that limit the transmission.

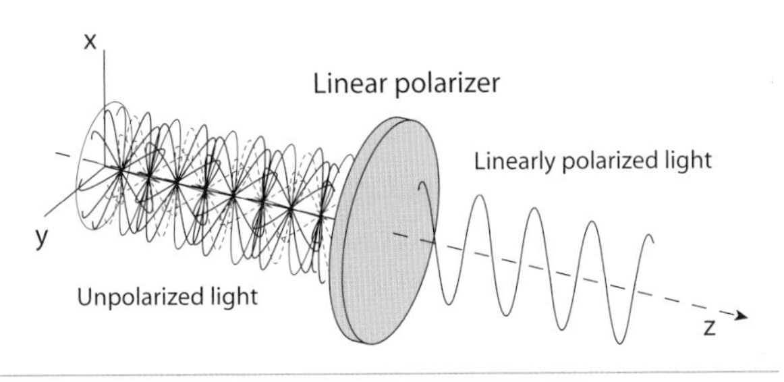

FIGURE 13.9. The interaction of unpolarized light and a linear polarizer.

Linear polarizers have been used successfully in photography for several decades, but, with modern cameras, problems sometimes arise from the interaction of polarized light with mirrors in the autofocus system and in the viewfinder. The efficiency of reflection depends on the state of polarization, and reliable operation may require unpolarized light. When it is desirable to take advantage of polarized light in a scene, there is a way to analyze for polarization and still insure that only scrambled polarization enters the camera. For this purpose one uses a circular polarizer in place of the linear polarizer. The operation of a circular polarizer is illustrated in Figure 13.10. A linear polarizer selects the direction of polarization and then the polarized beam passes through a quarter-wave plate. The result is circularly polarized light in which the electric-field vector rotates around the *z*-axis. The linear polarizer and the quarter-wave plate are actually glued together and act as a single filter. Within the camera, the circular polarized light acts like unpolarized light, and the problems vanish.

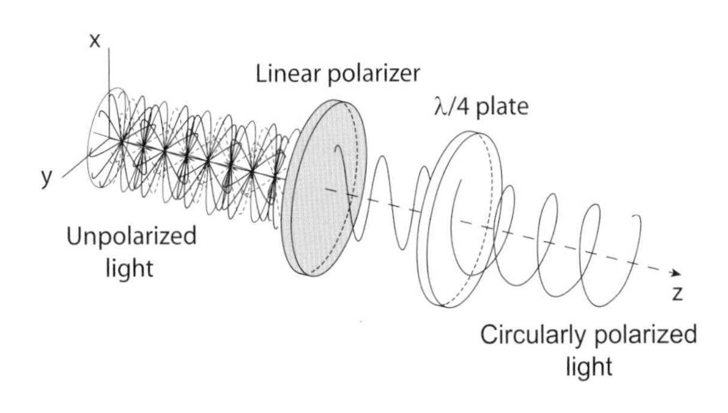

FIGURE 13.10. The interaction of an unpolarized beam of light with a circular polarizer.

So, what is a quarter-wave plate? Here we encounter doubly refracting (birefringent) crystals such as calcite. These crystals are anisotropic, and the index of refraction depends on the orientation of the polarization of the incident light. It turns out to be possible to cut some birefringent crystals into slabs (plates) so that the directions of maximum (slow axis) and minimum refractive index (fast axis) are perpendicular to each other and parallel to the surface of the slab.

The quarter-wave plate works like this: The birefringent plate is oriented so that the direction of polarization of the incident light lies between the fast and slow axes at 45° to each. As shown in Figure 13.11 the electric field E of the lightwave can be resolved into the components E_1 and E_2, which

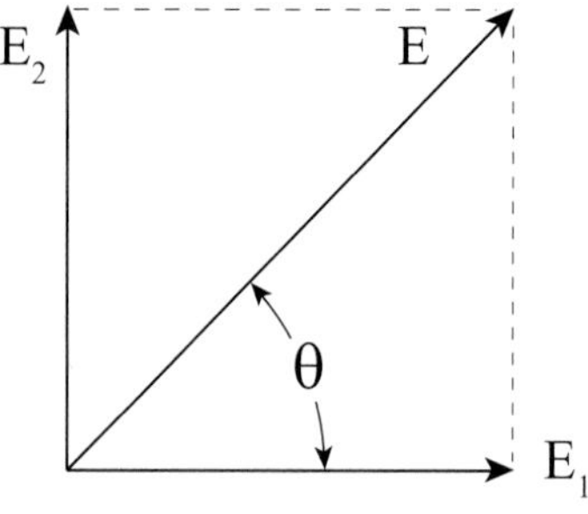

FIGURE 13.11. Resolution of polarized light into components.

lie in the directions of the fast and slow axes of the crystalline plate. Just as 2 + 2 is equivalent to 4, the sum of the component vectors E_1 and E_2 is equivalent to E. As discussed in Chapter 6, the apparent speed of light in a transparent medium with refractive index n is just c/n, where c is the speed of light in a vacuum, and the effective path length for light is the actual distance traveled multiplied by the refractive index. The quarter-wave plate is constructed by adjusting the thickness d so that the difference in path lengths for the components along the fast axis (refractive index n_1) and the slow axis (refractive index n_2) is one quarter wavelength or a 90° phase shift. This is best expressed with the equation $d \cdot (n_1 - n_2) = \lambda/4$, which can be solved for the thickness if the refractive indices and the wavelength are known. The result is that two waves are produced, in perpendicular planes, that have equal amplitudes and are shifted in phase by a quarter wavelength, and it can be shown that the sum of these two waves rotates around the z-axis as shown in Figure 13.10.[4]

It has been suggested that polarizing filters can be stacked to "increase" the polarizing effect, i.e., to narrow the distribution of light that is passed by the combination (from $\cos^2\theta$ to $\cos^4\theta$). This procedure will certainly fail for two circular polarization filters because the output of the first filter would be scrambled so that the second filter would not be able to improve the selection. Similarly, rotating one of the filters relative to the other would not change the fraction of light transmitted by the pair. The situation is different for a linear polarization filter in front of a circular polarization filter where the distribution is indeed narrowed (to $\cos^4\theta$). Also, rotation of one of the filters would attenuate the light completely when the filters were crossed. This combination could serve as a variable neutral-density filter. Two linear polarizers would behave the same way, except the output from the second filter would not be scrambled.

13.4 Polarization in Nature

Photographers are interested in polarizers because they want to be able to control naturally polarized light as it enters a camera. Polarized light usually results from reflection or scattering. I first consider scattered light. This is not a simple topic, but a simplified model captures all the important features, and we only need two ideas to explain the phenomenon:

1. The electric field of light exerts force on charged particles and, in particular, causes electrons in atoms and molecules to accelerate parallel to the direction of the field. These electrons oscillate at the frequency of the incident light wave.
2. According to Maxwell's electromagnetic theory, accelerating electrons emit radiation. This is well known from radio theory, where current oscillates in antennas and broadcasts a signal to listeners. (Recall that acceleration is the rate of change of velocity. This naturally happens in an oscillation where the velocity periodically changes direction.)

These ideas apply to cases where there is little or no absorption, and the scattered light has essentially the same frequency as the incident light. Figure 13.12 shows the angular distribution of radiation for an oscillating electron. The distance from the origin to the red curve at a particular angle indicates the amount of light intensity in that direction. The important points are that the direction of polarization is established by the direction of oscillation, and there is no intensity directly above or below the oscillator.

Consider, for example, a ray of sunlight in the sky well above the horizon. It is easy to verify with a polarizing filter that the light exhibits very little polarization when one looks toward the sun or

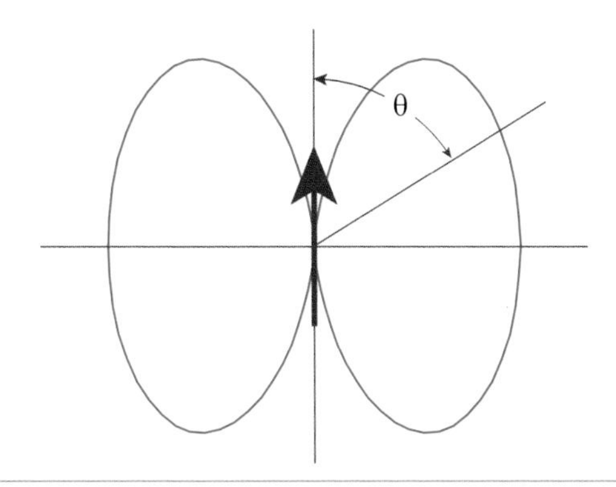

FIGURE 13.12. The radiation pattern for a dipole radiator.

away from the sun. Maximum polarization is at 90° from the direction of the sunbeam, and the sky becomes quite dark as the filter is rotated. Figure 13.13 illustrates this situation. Looking toward the sun, we receive radiation from electrons oscillating in all directions perpendicular to the beam direction. When the observer is in the plane of polarization, all electron oscillation appears to be in one direction. When viewed from 90°, the polarization may reach 75% to 85%. The polarization does not reach 100% because the oscillating electrons reside in molecules that are not perfect scatterers, i.e., the molecules are not perfectly spherical in shape and have a non-vanishing size, and, perhaps more importantly, some of the light is scattered by more than one electron before reaching the observer. Multiple scattering is the reason that light scattered near the horizon is unpolarized.

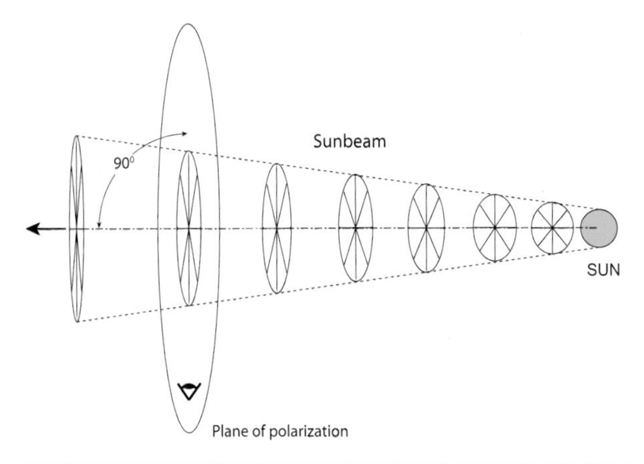

FIGURE 13.13. The transverse electric fields associated with unpolarized light from the sun.

Reflections are a major source of polarized light in nature. In the situations discussed here, light is reflected from a very smooth, nonconducting surface. (Surfaces that conduct electricity strongly absorb light and the treatment presented here does not apply.) In such cases, the fraction of the intensity reflected depends on both the angle of incidence and the direction of polarization of the light beam. Figure 13.14(a) illustrates the reflection and refraction of polarized light with transverse magnetic fields, which means that the electric fields are in the plane of the figure as indicated by the arrows (perpendicular to the scattering surface). In Figure 13.14(b), the incident beam is polarized with transverse electric fields, meaning that the electric fields are perpendicular to the figure surface as shown by the cross marks (parallel to the scattering surface). The indices of refraction above and below the reflecting surface are denoted by n_1 and n_2, respectively.

From previous discussions (Chapter 4), we know that the angle of incidence is equal to the angle of reflection ($\theta_1 = \theta_2$), and the angles of incidence and refraction are related by Snell's law ($n_1 \sin\theta_1 = n_2 \sin\theta_r$). The physical description of reflection and refraction is as follows: The electric field of the incident light beam penetrates the glass/water and induces electric currents (oscillating electrons). The currents produce both the reflected and the refracted beams and cancel the beam in the direction of the undeviated dotted line. Notice that the electric fields of the three beams (*I*, *a*, *b*) are all parallel in (b), and we expect that the currents will produce reflected light regardless of the angle of incidence. In contrast to this, the electric fields of the three beams (*I*, *A*, *B*) are not parallel in (a). Since the oscillating electrons move in the direction of the electric field and do not radiate in the direction of their motion (Figure 13.12), it is clear in the transverse magnetic case (a) that no light will be reflected when the refracted (*A*) and reflected (*B*) beams are perpendicular.

The remarkable conclusion is that when unpolarized light is reflected from a smooth surface at one particular angle, the components polarized

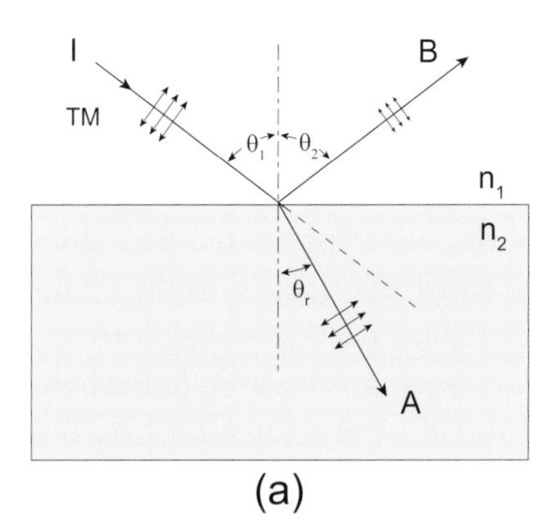

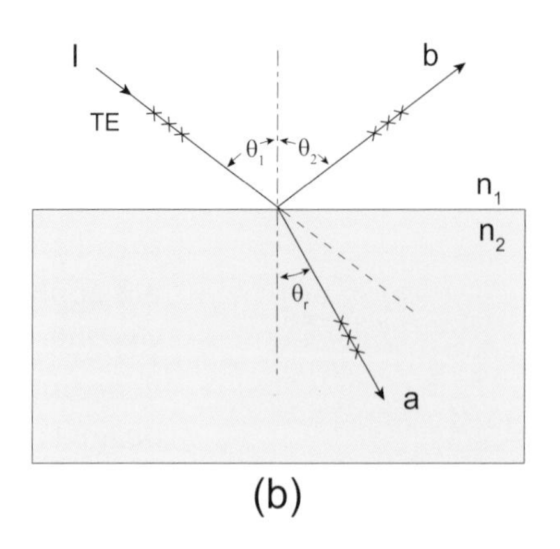

FIGURE 13.14. Reflection and refraction of light. (a) transverse magnetic field, and (b) transvers electric field.

perpendicular to the surface are not reflected at all. At this angle, known as the Brewster angle, the reflected light is polarized parallel to the surface, i.e., out of the plane of the figure. Brewster's angle can be determined by a simple calculation. When paths *A* and *B* in Figure 13.14(a) are perpendicular, the angle of the refracted beam is given by $\theta_r = 90° - \theta_1$. This expression can be combined with Snell's law and the identity sin $(90° - \theta_1) = \cos(\theta_1)$to obtain $n_2 / n_1 = \sin\theta_1 / \cos\theta_2 = \tan\theta_1$. The conclusion is that $\theta_1 = \tan^{-1}(n_2/n_1)$. For example, a light beam in air that reflects from the surface of water ($n = 1.33$) at the angle $\tan^{-1}(1.33) = 53.06°$, or 36.94°, measured from the reflecting surface, is totally polarized.

Of course, there is more to the problem than computing the Brewster angle. The percentage of reflected light from the TM and TE beams can be computed at all angles of incidence. This can be done by making use of simple geometric arguments or by applying Maxwell's equations with appropriate boundary conditions to describe the electric and magnetic fields. The Maxwell equation approach, while more demanding, does permit reflection from absorbing material to be treated. Since these calculations would lead us too far afield, I simply present the results and provide references for the complete derivations. The fraction of light reflected for TM and TE arrangements are described by Equations (13.1):

$$R_{TE} = \left(\frac{\sin(\theta_1 - \theta_r)}{\sin(\theta_1 + \theta_r)}\right)^2, \; R_{TM} = \left(\frac{\tan(\theta_1 - \theta_r)}{\tan(\theta_1 + \theta_r)}\right)^2. \quad (13.1)$$

These quantities are plotted in Figure 13.15 for the special case with $n_1 = 1$ and $n_2 = 1.33$.

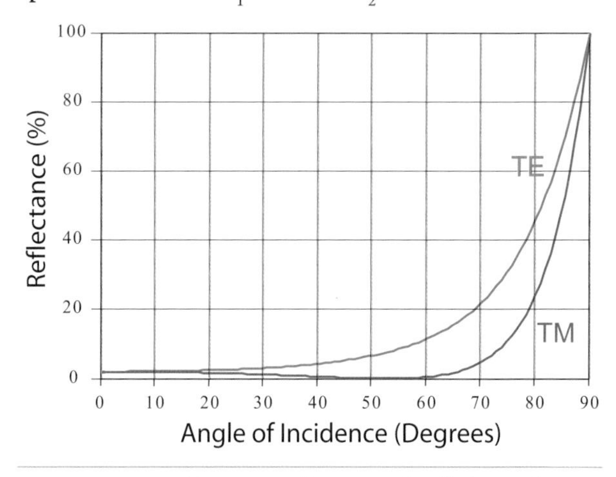

FIGURE 13.15. Percent reflectance for light at the air–water interface.

Polarization (TM/TE) at the Brewster angle (53.06°) is clearly shown, but it is also evident that polarization is strong from at least $\theta_1 = 40°$ to 60°. The results are slightly different for glass where $n_2 \approx 1.5$. In summary, at glancing incidence ($\theta \approx 90°$), reflectance is close to 100% for both TE and TM components, and the reflected beam is not polarized; at normal incidence ($\theta = 0$), about 4% is reflected with no polarization. Between glancing and normal angles there is some polarization, and it becomes complete at the Brewster angle.

Polarization in a rainbow. Rainbows are favorite subjects for photographers, but few realize that the rainbow light is polarized. First, recall that rainbows are circles, and their center is always the shadow of

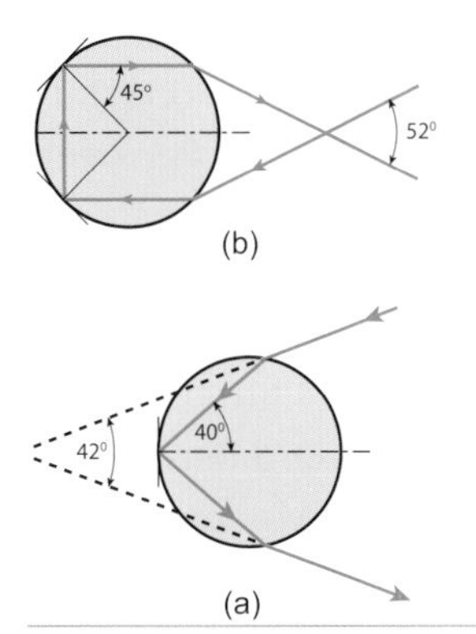

FIGURE 13.16. Refraction and reflection of light beams in rainbows: (a) primary and (b) secondary.

the head of the observer, i.e., the anti-solar point. It turns out that the light from the rainbow is polarized tangent to the colored bands or perpendicular to a radial line from the center (see the yellow arrows in Figure 13.16). This happens because sunlight is reflected inside a water droplet and the droplet lies in the plane containing the droplet, the observer's head and the shadow of the observer's head. Furthermore, the angle is close to the Brewster angle for internal reflection. The geometry of scattering is shown in Figure 13.16. Colors result from the dispersion in the index of refraction of water that ranges from 1.344 at 400 nm (blue) to 1.3309 at 700 nm (red), but reflection (inside water droplets) is responsible for the polarization.

A study of the possible angles for internal reflection, when there is only one reflection inside a droplet, shows that the maximum angle (from the anti-solar point) is 40.5° for 400 nm light and 42.4° for 700 nm light. This gives the primary rainbow where the internal angle of reflection is about 40° and the Brewster angle is close to 37°. The secondary rainbow results from two internal reflections, and I find that the minimum angle for 400 nm light is 50.34° and for 700 nm is 53.74°. In principle, the secondary rainbow has about 43% of the intensity of the primary, but it is almost invisible in Figure 13.16. The internal reflection angle for the secondary rainbow is about 45°, not quite as close to the Brewster angle as with the single reflection. The consequence of the limiting reflection angles is that the colors are reversed in the secondary rainbow going outward from red to blue, and that no light is reflected between roughly 42° and 51° by either single or double reflection. This gives rise to the dark band between the primary and secondary rainbows that was described by Alexander of Aphrodisias in 200 CE and bears his name. Additional discussion and illustrations can be found in the book by Lynch and Livingston.

Reflection by metals. So far I have limited the discussion to reflection from nonconductors. The problem with conductors of electricity is that they absorb light and, in some cases, absorb it very strongly. Actually, the stronger the absorption the higher the reflectivity, and the reflectance can reach nearly 100% without any polarization for polished metals. More poorly reflecting metals such as steel partially polarize reflected light. For a typical metal, reflectance of both TM and TE beams depends on the angle of incidence, and the TM beam shows minimum, though nonzero reflectivity, at the principle angle of incidence. There are no universal polarization curves for metals since the reflection of polarized light depends on the optical absorption coefficient, which in turn depends on the wavelength of the light. Photographers must depend on trial-and-error methods with metals.

13.5 UV and IR Photography

The human eye can only see a very narrow band of radiation between about 400 nm and 700 nm that is defined as the visible region. Thus we are restricted to a region of the electromagnetic radiation spectrum where the frequency varies by only a factor of 2. At first glance, this seems remarkable since the complete spectrum from gamma rays to radio waves encompasses almost 15 orders of magnitude change in frequency. However, humans evolved the ability to see sunlight, and most of the energy from the sun lies in the visible region. The solar flux is illustrated in Figure 13.17. At wavelengths less than 400 nm, where the UV-A region begins, the intensity of the solar spectrum decreases, and it is severely attenuated below about 320 nm because of absorption by ozone. This loss of intensity is a very good thing, since photons in the UV region have high energy and can damage skin cells as well as the photoreceptors in the eye. In the infrared region, beyond 700 nm, there is still considerable intensity, but the photons have low energy, and the photoreceptors in the eye are not very efficient at their detection.

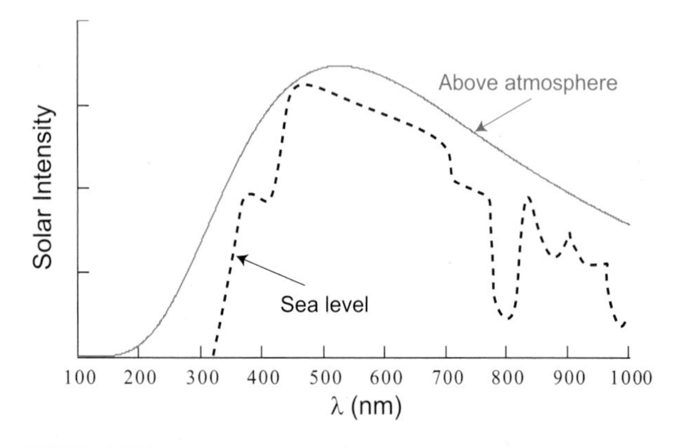

FIGURE 13.17. Sketch of solar intensity versus wavelength above the atmosphere and at sea level.

So we accept the fact that human vision is not good in the UV and IR regions, but what about photography? First consider the near UV region (320–400 nm, UV-A). The major problem we encounter is that glass in standard lenses begins to absorb light at about 420 nm and by 370 nm transmits very little. Practitioners in UV photography use special quartz/fluorite lenses that transmit UV at least to 300 nm. The UV-Nikkor 105 mm lens is often mentioned, but it is hard to find and is extremely expensive. The next thing that is necessary for UV photography is a filter that can completely block visible light. The Kodak Wratten 18A transmits from 300 nm to 400 nm and blocks visible light, but it also transmits IR from 700 nm to 900 nm. Other UV transmission filters, such as the B+W 403, which are less expensive, have the same IR problem. The way to obtain only UV light is to combine a Wratten 18A or similar filter with a hot-mirror IR reflective filter. This is certainly possible, and a number of examples have been reported. Finally, one needs a sensor that can detect UV. The CCD sensors in digital cameras are not very sensitive in the UV but are useable out to perhaps 350 nm. It is even possible to obtain interesting results with some standard lenses, and experimentation is recommended. Modern multi-element multicoated lenses are likely to be the worst choice because there are more lens elements to absorb UV. As an example, I show images obtained with a Canon G-1 camera with and without a B+W 403 UV transmitting filter (Figure 13.18). As noted, the filtered image is contaminated with IR radiation. The UV image was brightened with Photoshop, but the false colors were a surprise.

Fortunately, IR photography is fairly easy and often rewarding for amateurs. Both color and B&W films with IR sensitivity are available from Kodak (Kodak Ektachrome Professional Infrared EIR Film, 380 to 900 nm, and Kodak HIE High-Speed Infrared, to 900 nm); and both Ilford and Agfa have

FIGURE 13.18. Paul Green Cabin. Canon G-1 camera (a) without filter and (b) with B+W 403 UV-pass filter.

advertised B&W films that are sensitive to nearly 800 nm. Also, lenses are not a big problem in IR photography. Many, but not all, standard lenses are satisfactory for IR photography. The problem, when it arises, is that some lenses give "hot spots" in IR photographs. One can select lenses by trial and error or by looking online for lists of lenses that have been found to be satisfactory.

Of course, filters have to be used to block visible light, and here there are a lot of choices. One has to decide on the desired cutoff point. Some popular filters are listed in Table 13.2. Many other filters are available, and equivalent filters can often be obtained from B+W, CoCam, Cokin, Heliopan, Hoya, Schott-Glass, and Tiffen.

TABLE 13.2.
Red and IR filters

Kodak Wratten filter	25	89B	87	87c
0% Transmission (nm)	580	680	740	790
50% transmission (nm)	600	720	795	850

In digital photography, we must deal with CCD or CMOS sensors in place of film. These sensors are inherently sensitive in the red/IR region—so much so that without blocking ambient IR the recorded colors may be distorted, i.e., IR will be falsely recorded as a visible color. Therefore, manufacturers have installed hot-mirror IR-reflectors to prevent IR radiation from reaching the sensors. Of course, the hot mirrors are not perfect, and some IR gets through. So we can add a filter to block visible light and hope that enough IR gets through the hot mirror to permit an image to be recorded with a reasonable exposure time. As you might guess, some cameras are accidentally better than others at producing IR images this way. As an example, I used a Canon G-1 (3 megapixel) with a Hoya 72R (Wratten 89B equivalent) on a tripod to photograph Mabry's Mill on the Blue Ridge Parkway in Virginia). Figure 13.19 shows this photograph compared to one taken (handheld) without the IR filter. Both were converted to monochrome

FIGURE 13.19. Mabry's Mill. Canon G-1 camera (a) without a filter, f/4.0, 1/160 s, and (b) with a Hoya R72 IR filter, f/5.6, 1.0 s.

in Photoshop. The obvious differences are in the foliage, which is white in the IR photo, and the water, which is black in the IR photo. Blue sky would also be very dark in an IR image, but in this case there were thin clouds. In general, IR images are more striking than ordinary B&W, and it is possible to obtain very interesting false-color images. The exposure time was very long for the IR shot, but the experience was not bad because the image was clearly visible on the camera's LCD screen even though the filter looked quite opaque.

Of course, action photos are out of the question with long exposures, and even a breeze can upset things by moving leaves and branches. The way around this problem is to have the IR filter (reflecting hot mirror) removed from the digital camera. Several companies offer this service now for both point-and-shoot cameras and DSLRs. There are two options. The IR filter can simply be removed, and the user then must use an external filter for normal photography and a visible blocking filter for IR photography. The down side of this arrangement for DSLRs is that one cannot see through the dark IR filter and must remove the filter to compose the scene. The other option is to have the IR blocking filter (hot-mirror) in the camera replaced with a visible blocking filter mounted directly on the sensor. With this arrangement, the DSLR works normally and is convenient for action photography. The IR conversion is an attractive option for an older DSLR that has been replaced with a newer model. Companies that advertise IR conversions include MaxMax and Life Pixel.

Further Reading

G. R. Fowles. *Introduction to Modern Optics, Second Edition*. New York: Dover, 1989.

E. H. Land. "Some Aspects of the Development of Sheet Polarizers." *J. Opt. Soc. Am.* 41 (1951), 957–962. (Dr. Land discusses the history of polarizers.)

D. K. Lynch and W. Livingston. *Color and Light in Nature, Second Edition.* Cambridge, UK: Cambridge University Press, 2001. (This is a beautiful and fascinating book.)

R. P. Feynman, R. B. Leighton, and M. Sand. *The Feynman Lectures on Physics.* Reading, MA: Addison-Wesley, 1963. (This is a discussion of polarization for students of physics. See Vol I, Section 33–3; Vol. II, Chapter 33.)

CHAPTER 14

The Limits of Human Vision

How Good Does a Photographic Image Need to Be?

The eye is the jewel of the body.

—HENRY DAVID THOREAU

You can observe a lot by watching.

—YOGI BERRA

14.1 Introduction

Photography is a human activity that depends entirely on the capabilities and limitations of the human visual system. We judge the apparent resolution, brightness, contrast, and the quality of color in photographs, but these attributes are subjective since everything depends on perception. Our vision results from a combination of physiology and psychology that correspond roughly to the eye and the brain. Of course, all eyes are not the same, but a typical or standard observer must be assumed to exist by those designing photographic equipment. Therefore, it is necessary to consider what the typical healthy eye can see. A study of the eye can provide us with criteria for image resolution and color stimuli that humans can distinguish and appreciate in photographic images. The perception of color, resolution, and overall image quality depends, of course, on the wiring of nerve cells in the retina and the visual cortex. Consideration of this part of the visual system can be found in Chapters 15 (color), 17 (resolution), and 18 (art appreciation).

14.2 Structure of the Eye

First, we look at the structure of the eye. The eye has some of the features of a camera, or better yet a TV camera. The relevant components are shown in Figure 14.1. On the left are the cornea and the lens, which work together as a compound lens. Most of the focusing power (about 40 diopters) results from refraction at the air and cornea interface, while the lens provides adaptive focus or accommodation over a range of about 20 to 30 diopters. With a relaxed lens, the focal length of the combination is about 22 mm, taking into account the refractive index of the vitreous humor ($n = 1.33$). Immediately in front of the lens, we find the iris or diaphragm, which controls the size of the pupil. The iris is the colored part of the eye, while the pupil appears to be black. The pupil ranges in diameter from less than 2 mm to more than 8 mm, so that the optical system can be described as about $f/10$ to $f/2$.

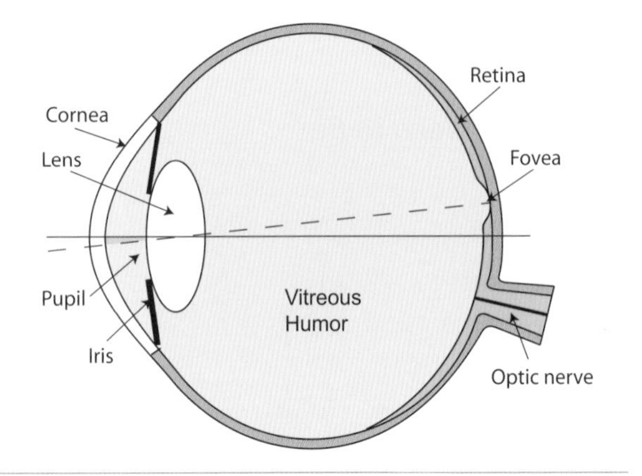

FIGURE 14.1. Simplified schematic of the eye.

The projected image falls on the retina shown at the right in Figure 14.1. The retina is a structure about 0.5 mm thick that covers the back of the eye and contains the photoreceptors. The top layer (facing the lens) is made up of ganglion, or output neutrons, the middle region consists of nerve tissue, and the bottom layer contains rods and cones (the visual photoreceptors). The central layers of nerve tissue are quite complex and contain networks of horizontal cells, bipolar cells, and amacrine cells connecting the photoreceptors to the ganglion cells. Behind the photoreceptors is a black layer, the pigment epithelium, that absorbs light and limits back reflection. A schematic of structure of the retina is shown in Figure 14.2. Overall, the structure appears to be upside down. Light has to pass through multiple layers, where it is attenuated and dispersed before it reaches the receptors. This arrangement is certainly expected to limit visual resolution. The structure of the retina is not uniform, however, and there is a special region (described in Section 14.3) that permits high acuity (visual sharpness).

In addition to the well-known rods and cones, it has been discovered that that there is a fifth type of photoreceptor in the eye. It turns out that some

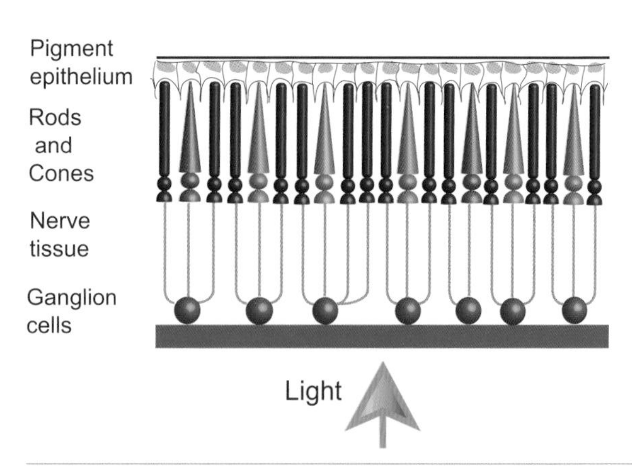

FIGURE 14.2. Schematic diagram of the retina.

of the ganglion cells are photoreceptors as well as neurons and are part of an independent system that monitors the level of illumination for control of physiological functions including phasing of the circadian clock and control of pupil size. These ganglion cells respond to blue light (460–500 nm) and may make use of melanopsin as the photopigment. However, this system is apparently not directly related to vision and is not relevant to our consideration of the eye as an optical instrument (see [Hattar et al. 02]).

14.3 Operation of the Visual Receptors

To understand how the eye accomplishes high resolution and color vision as well, one needs to consider the visual receptors. There are about 130 million receptors in the retina. Of these, about 120 million are rods that are highly sensitive receptors containing the visual pigment rhodopsin. A rod can detect a single photon. The rods respond to blue/green light and have maximum sensitivity at about 500 nm. They are responsible for night vision but permit no resolution of color. In addition, there are 6 or 7 million cones, which contain visual pigments called *opsins*. Experiments have demonstrated that cones come in three varieties, commonly known as red, green, and blue cones. Actually, each type of cone responds to a wide range of frequencies, but their frequencies of maximum absorption are shifted as follows: 564 nm (red or long wavelength L-cones), 534 nm (green or medium wavelength M-cones) and 420 nm (blue or short wavelength S-cones). The normalized absorption spectra for human photoreceptors are shown in Figure 14.3.

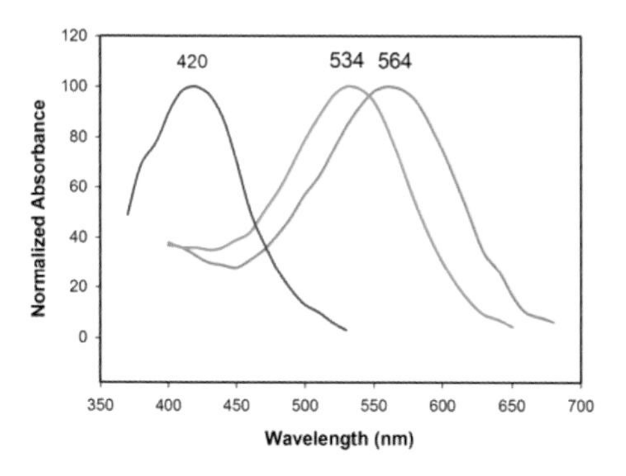

FIGURE 14.3. Absorption spectra of human cones (adapted from [Bowmaker and Dardnall 80]).

It turns out that our visual acuity and color vision depend on a special region of the retina, known as the *fovea centralis* or *fovea* for short, that is about 0.4 mm (400 μm) in diameter. The entire region responsible for high-acuity vision including the foveal pit and the foveal slope (perifoveal and parafoveal regions) is known as the macula region. In illustrations of the retina, the fovea is usually shown as a notch or pit in the surface because the neuron layers are pushed out of the way in this small area to expose the photoreceptors to undisturbed photons from the lens. This structure is illustrated in Figure 14.4. The foveola contains 30,000 cones in a tight mosaic pattern and no rods. In the foveola, the cones have diameters of from 1 to 4 μm, while elsewhere in the retina, where their density is much lower, their diameters are 4 to 10 μm. The rods are distributed throughout the retina, but their maximum density is found at about 4 to 5 mm from the fovea. Both rods and cones are more or less rod-shaped with lengths of about 50 μm perpendicular to the surface of the retina. Another way the fovea gets special treatment is that the ratio of cones to ganglion cells is only 3 : 1 while elsewhere in the retina the ratio of receptors to output neurons is 125 : 1.

The rods are much more sensitive to light than the cones, but in bright light they are saturated

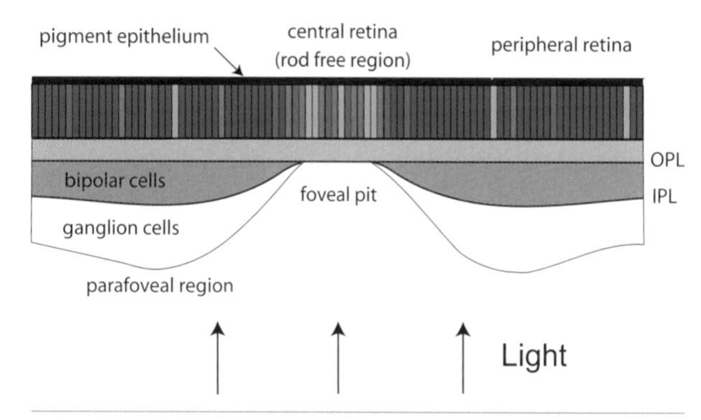

FIGURE 14.4. Schematic of the macula region; the rods are shown in purple and the RGB segments are cones.

and do not function. In dim light, the cones are not effective, but as light intensity increases the cones take over and, furthermore, provide information about the distribution of wavelengths in the stimulus that permits colors to be distinguished. Surprisingly, the three types of cones are not present in anything close to equal amounts in the fovea. The blue (S) cones, which are somewhat larger in diameter and more sensitive than the red (L) or green (M) cones, only comprise about 3% of the cones in the fovea. How the excitation of these cones leads to color vision is an interesting but incompletely understood subject, that I will return to in Chapter 15.

14.4 Visual Acuity

The acuity of the eye as an optical instrument is the easiest part to understand. As shown in Figure 14.4, the fovea, with a diameter of about 0.4 mm, subtends an arc of about 1° as required by the equation, arc length = (focal length) × angle, where the focal length is 22.3 mm. This represents the angular range for high-acuity vision and is about twice the angle subtended by the sun or moon. This foveal diameter also corresponds to the area of 30,000 cones each having a radius of about 2 μm. Since the resolution of points or closely spaced lines depends on cones whose centers are separated by about 4 μm in the fovea, the next question is what angle of view corresponds to this spacing. By using 4 μm as the arc length of a cone, we find an angle of about 0.6' of arc. This is close to the value of 1/60° = 1 ft. that is usually assumed for visual acuity in photography and for 20/20 vision.[1] To put this in perspective, I note that a dime (18 mm) at 200 ft. subtends an angle of 1 ft.

Of course, the question of resolving power for the eye or any other lens system is more complex than I have described. Suppose that light comes from a point source, say a star, and the angle subtended is close to zero. The image projected on the retina has a width that depends on the size of the pupil and aberrations in the cornea and lens. When the aperture (pupil) is 1 mm, the spot size and intensity pattern are determined primarily by diffraction as in a pin-hole camera. For large pupil sizes, aberrations dominate, and the optimum pupil size for resolution is between 3 and 5 mm. In general, the width of the point-spread function (PSF) indicates the quality of the optical system. For straight-line light sources, the line-spread function (LSF) gives similar information.

Another way to explore resolving power is to measure the eye's response to a pattern of equally spaced white and black lines as a function of their spacing, i.e., line pairs per degree (lpd). This type of test can also be performed with a sinusoidal pattern of white peaks (maxima) and black valleys (minima), in which case the spatial frequency is denoted by cycles per degree (cpd). The idea is to measure the contrast of the image on the retina and to compare that with the contrast of the original object pattern. If the peaks and valleys in the image have the intensities I_{max} and I_{min}, respectively, the contrast of the image is defined as

$$C_i = \left[\frac{I_{\max} - I_{\min}}{I_{\max} + I_{\min}} \right].$$

The ratio of the image contrast C_i to that of the original stimulus C_0 then defines the modulation transfer function for the particular spatial frequency:

$$\text{MTF}(\nu) = C_i / C_0,$$

where ν denotes the spatial frequency in line pairs or cycles per length or degree. The function MFT(ν) is mathematically related to the PSF. What we find is that the apparent contrast decreases as the spatial frequency increases until only a uniform gray with zero contrast can be seen. This procedure is illustrated by the stimuli and responses in Figure 14.5.

The MTF is often used as a measure of optical quality, and the frequency at which the MTF drops to 0.5 is taken as an indicator of lens resolution. It definitely provides information about acuity, but it does not give complete information about phases of light rays. The MTF for the human eye at various pupil sizes is shown in Figure 14.6. These curves were computed with the equations reported in [Rovamo et al. 98]. It was concluded that the eye is close to a diffraction-limited optical system at a pupil size of 1 mm but is considerably worse than diffraction-limited at larger pupil sizes.

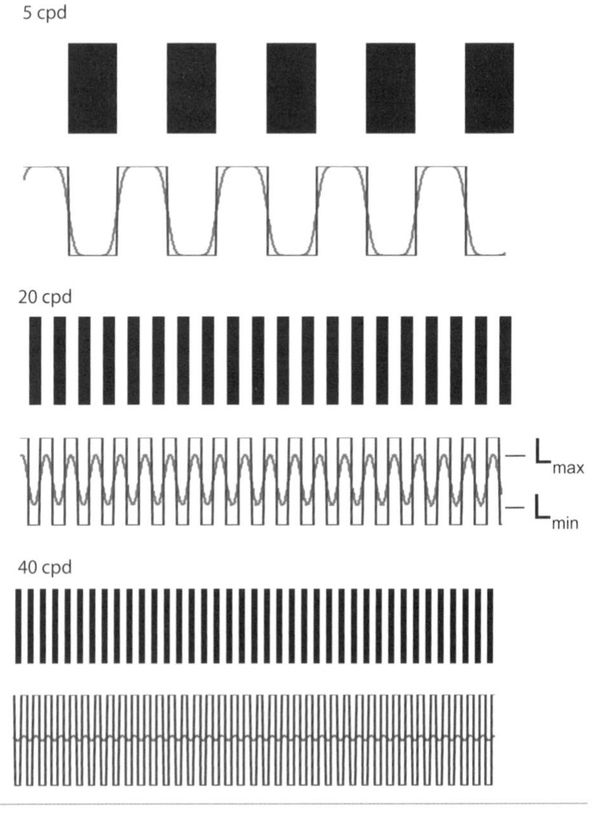

FIGURE 14.5. Stimuli (black) and responses (red).

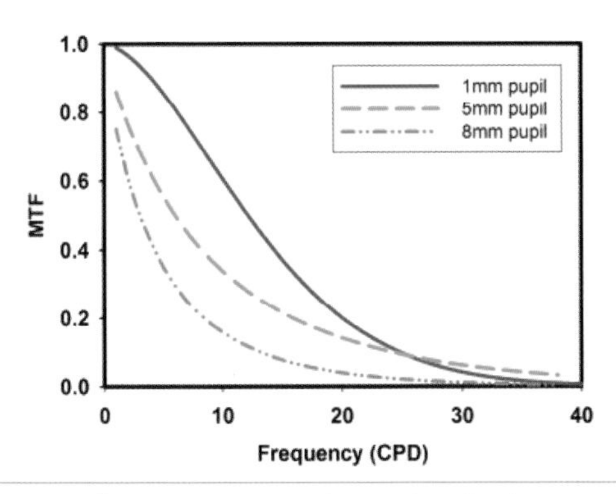

FIGURE 14.6. Foveal optical MTF for the human eye for various pupil sizes.

It is important to point out that Figure 14.6 describes the optical system of the eye and not the MTF for the complete visual system. Psychophysical experiments reveal that humans have most sensitive response to frequencies close to 6 cpd and that response depends strongly to contrast. When contrast sensitivity is taken into account, the complete MTF has a maximum close to 6 cpd and falls off at both high and low frequencies. The high-frequency region is dominated by the optical properties that are responsible for the curve in Figure 14.6, while the low-frequency part depends on the properties of frequency-sensitive cells in the visual cortex. (For a more complete discussion of this point, see Chapter 17.)

A comparison of the resolution of the eye and the digital camera also involves estimation of the effective number of photoreceptors (pixels) in the eye (see [Clark 09a] for interesting insights and neat calculations). We have seen that the fovea subtends only about 1 deg of arc in the field of view (FOV), but that it seems to provide visual acuity everywhere by scanning as needed. This is somewhat analogous to a high-resolution film scanner with only a single row of detectors, but which captures details from an entire negative by scanning. We ask now how many cones (sensors) would be required in the retina for the eye to be able to capture the complete FOV simultaneously as electronic sensors do. This calculation is easy after we decide what the FOV is. The eye has a total horizontal FOV of at least 180°, but our impression of the high-resolution field is much narrower. I am going to assume arbitrarily 120° and 80°, for the horizontal and vertical FOVs, respectively. This angular rectangle corresponds to a solid angle of 2.4 steradians (sr), while the foveola subtends only 2.4×10^{-4} sr. (As a reference, 4π is the solid angle for all space; please see ["Solid Angle" 09] for an explanation and to find more about the calculation of solid angles.) Therefore, the rectangular FOV has an area of about 10,000 foveas, and, since each fovea contains 30,000 cones, we conclude that the number of effective pixels is 300 million. So digital cameras have a way to go, but they do provide detection of color at low light levels, no delay for dark adaptation, and, of course, the ability to transfer and store images.

14.5 Sensitivity

The sensitivity of the eye is another property that can be compared with a digital camera. In bright light, the cones are effective (photopic vision), and at very low light levels the rods take over (scotopic vision), giving much higher sensitivity. The effect is something like changing the ISO sensitivity of a camera sensor except that dark adaptation of the eye can require up to 30 minutes. Clark estimated the ISO equivalent of the dark-adapted eye by comparing the ability of the eye to see magnitude 14 stars through a 5 in. aperture lens with a Canon 10D camera in the same situation, and he concluded that the eye offers about ISO 800 sensitivity. In bright light, where the rods saturate and the cones take over, sensitivity has been estimated to be about ISO 1. An important difference between the retina of the eye and the sensor in the camera is that the eye always has low-sensitivity cones and high-sensitivity rods in place, while the camera sensor resets all detectors to the same ISO sensitivity. As a consequence, even the dark-adapted eye has the high acuity fovea region totally populated with low-sensitivity detectors (cones). This design feature of the eye explains why some faint stars can be seen only by moving the center of vision a few degrees away from the star so that the star image falls on rods rather than cones.

There is a tendency to compare rods in the human eye with detectors in a sensor, and to ask if a rod can detect a single photon. The answer appears to be yes, but the observer would not be aware of the event. A conscious response requires the rods to detect several photons in a short period of time. The threshold for detecting light falling on 10 min. of arc in the region of highest concentration of rods on the retina is in the range of 10 to 100 photons. Considering the number of rods involved, the rods must be responding to single photons. This experiment emphasizes the fact that in the visual system everything depends on the wiring of the photoreceptors and the firing of nerve cells.

The impressive thing about the visual system is how the signals from 130 million receptors in two eyes can be processed to give useful images in real time. This brings us to questions about how signal processing is distributed between the retina and the visual cortex in the brain and basically how the brain works. While that is beyond the scope of this book, I discuss the major theories of color vision in the next chapter and ideas about visual perception in Chapter 18.

14.6 Conclusions

Furthermore, the cones provide three color receptors and justify to some extent our use of three colors to generate all the colors in color displays. However, as we shall see, it is not possible to synthesize all the colors we can perceive by mixing light from three narrow band (monochromatic) sources. Similarly, our electronic sensors do not have the same distribution of sensitivities as human cones. Detection and display systems based on three primary colors can provide a good, but not perfect, representation of the colors our eyes respond to. Obtaining an image that matches the appearance of the original scene is a more challenging problem. The trichromacy color theory is discussed in Chapter 15 along with the alternative opponency theory. Other aspects of the visual system are presented in Chapters 17 and 18.

Further Reading

D. H. Hubel. *Eye, Brain, and Vision,* Scientific American Library, 22. New York: W. H. Freeman, 1995. Available online at http://hubel.med.harvard.edu/bcontex.htm..

B. A. Wandell. *Foundations of Vision.* Sunderland, MA: Sinauer Associates, 1995.

H. Kolb, E. Fernandez, and R. Nelson. Webvision: The Organization of the Retina and Visual System. Available at http://webvision.med.utah.edu/, 2009. (Great overview and wonderful illustrations. In particular, see "Visual Acuity" at http://webvision.med.utah.edu/Kallspatial.html, which has everything about acuity including eye charts and intensity effects.)

J. T. Fulton. "The Standardized Human Eye." Available at http://www.4colorvision.com/pdf/standeye.pdf, 2004. (Every fact you can imagine about the physical properties of the eye.)

CHAPTER 15

How Can Color Be Managed?

For the newly sighted, vision is pure sensation unencumbered by meaning.

—ANNIE DILLARD

But, to determine more absolutely, what Light is, after what manner refracted, and by what modes or actions it produceth in our minds the Phantasms of Colours, is not so easie. And I shall not mingle conjectures with certainties.

—ISAAC NEWTON

15.1 Introduction

Color vision is possible because three types of cones (S, M, and L) in the retina respond to slightly different ranges of wavelengths of light. This part is well understood in terms of chemistry and physics, but it only represents the first steps in a truly amazing process. The perception of color and, in fact, the complete illusion of a multicolored world, is generated in the visual cortex of the brain. Information about the distribution of intensities at various wavelengths available from the three types of photoreceptors is first processed by bipolar and ganglion cells in the retina and is then transmitted via axons in the optic nerve to the *lateral geniculate nucleus* (LGN). In the LGN, signals from the two eyes are integrated and projected directly to the *primary visual cortex* (V1), that occupies the surface of the occipital lobe of the cerebral cortex (see Figure 15.1).

In the last few decades of the 20th century, noninvasive imaging methods became available for the study of normal, healthy human brains. In particular, functional magnetic resonance imaging (fMRI) permits brain response to visual stimuli to be correlated with specific locations in the visual cortex. These developments have prompted a large number of studies of localized brain function. The consensus view is that the visual cortex contains a set of distinct visual field maps, and that nearby neurons analyze nearby points in the visual field. While the cells in V1 appear to respond to wavelength, the perception of color requires additional processing. Visual maps have been established in the visual cortex areas V1, V2, and V3; 20–30 visual areas have been distinguished, and the following picture has emerged. Information is transmitted via two pathways. In the *dorsal stream,* information passes from V1 through V2 and on to the middle temporal area (MT or V5). This is the "where pathway" that is concerned with location and motion of objects and control of the eyes. The *ventral stream*, or "what pathway," also begins with V1 and V2, but directs information to the visual area known as the V4 complex and the adjacent V8 area. This path is involved with representation, long term memory, and the association of color with visual stimuli.

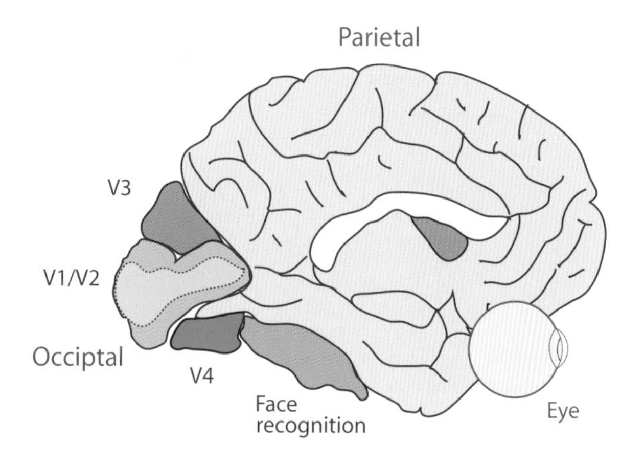

FIGURE 15.1. A sketch of the human brain showing the lobes (blue type), the primary visual cortex (V1), and other specialized visual areas as described in [Zeki 99].

The exact location of visual processing is not the primary concern here, but rather the fact that neural processes deep in the visual cortex of the brain are the origin of color. Objects don't have color, and neither do light beams. Objects simply absorb, transmit, and reflect various amounts of incident electromagnetic radiation, and the radiation is characterized by wavelength and intensity. If radiation falls in the range of about 400 nm to 700 nm, the photoreceptors in our eyes can respond to it (see Figure 14.3). The response is processed and used in various ways to analyze the visual field for patterns and motion, but the association of color with the excitation of cones in the retina is another thing altogether. Part of this picture has been revealed through the study of defects in color vision. Color vision deficiency (dyschromatopsia) is often genetic in nature: one or more of the three types of cones (S, M, or L) may be missing or defective. Thus about 1 in 12 males and a smaller fraction of females suffer from some color vision deficiency, the most common of which is red–green color-blindness. Each deviation in function of a cone generates a different perception of the world around us, and it is clear that a shift in the region of maximum sensitivity of all the cones and rods by a few hundred nm would present to us a strange, alien world with little similarity to the one we now accept as reality.

Now, suppose that color perception is destroyed by an injury or a disease that affects visual areas in the cortex, perhaps the V4 complex. Such events are, in fact, well documented, and the consequences are more dire than might be imagined from the simple loss of color for the visual field. Cerebral achromatopsia patients experience a drab and disturbing world with little similarity to black-and-white movies or TV. In his book *An Anthropologist on Mars*, Oliver Sacks described the case of an artist (Jonathan I.), who suffered from achromatopsia after an automobile accident [Sacks 95]. This was a catastrophic loss for an artist, who had previously produced brilliantly colored paintings. He found that "... now color was gone, not only in perception, but in imagination and memory as well." There was a "wrongness" to things, and photographers will be able to relate to the suggested analogy with "Tri-X pushed for speed." However, Mr. I. strongly maintains that the concept of gray does not at all capture the shades he experiences.

Color perception, and, indeed, all vision, results from complex multichanneled information processing in the brain. We must not, however, imagine a unidirectional information flow with data from photoreceptors being processed at ever more-sophisticated levels and finally emerging as evolving color images for contemplation and reaction. Instead, there is a high degree of feedback at all levels so that what we see depends on more than our conscious control of attention. Priority is given to what we must see for survival. A snake or lion receives our immediate attention and initiates action before we can consciously consider consequences. Also, for color to serve as a biological marker, there must be some constancy in the apparent colors of objects even when they are illuminated by quite different light sources. This implies a kind of automatic white balance adjustment. In addition, to some extent we see what we expect to see. Fleeting images, especially of poorly illuminated objects, often take on appearances of well-known things, e.g., roots appear to be snakes.

Furthermore, those who deal with images need to recognize that vision with comprehension is a learned process. A normal visual system presents an immense array of patterns and colors to a newborn baby, but initially there is no association of this information with reality. Babies have no perception of depth, and small children only slowly learn how to put round objects in round holes, square things in square holes, etc. The child learns to use his or her eyes efficiently, to experience depth, to recognize shapes, to respond to illusions, etc. It has, in fact, been estimated that the complete development of vision in man requires up to 15 years. Eventually, a model of reality is learned so that an individual is comfortable with their location in space, as revealed by the eyes. However, adults who have matured in restricted environments, e.g., as members of tribes living in tropical rain forests may still have limited depth perception and may fail to recognize two-dimensional pictures as representations of the three-dimensional world. Here again our understanding of cognition and vision has been greatly extended by case studies of individuals who have regained sight, after extended periods of blindness through cataract operations. A newly sighted individual has to learn to see, and the transition from blindness to sight is fraught with difficulties. There even appears to be mental conflict in adapting a brain organized for blindness, with enhanced touch and hearing, to a brain organized for seeing. The starting point for a baby is quite different. Learning may still require an extended period, but with a baby there is little internal conflict.

I want to emphasize that each of us has a unique visual system both in capabilities and conditioning. We may achieve similar scores on visual acuity tests, but still perceive quite different worlds. Even those who show no visual dysfunction may not see similar colors. Language, of course, is no help. We all use the same name for the color of an object, but no one can know what others see. This brings us to a dichotomy that can serve as the theme for the study of visual perception and photography. Differences in perception do not pose a severe problem for the design of color cameras, monitors, printers, etc., because the goal is to replicate in an image the optical properties of objects in a framed scene. What our eyes and brain do with the light rays from an object or its photographic image is outside the realm of color management in photography. As we shall see later in this chapter, the theory of light-mixing in photography, trichromatism (George Palmer, 1777; Thomas Young, 1802), depends on viewing a small visual field with a dark background (the void mode). The aim of color management in photography is for each point (pixel) in the image to reproduce the visual characteristics of a point on an object. To go beyond this, and to control appearance as well, one must be able to specify viewing conditions.

In fact, our perception of colors in a scene and in a color photograph of the scene depends on the neighboring colors and their brightness. Optical illusions also depend on image elements in context. While composition and the interactions of image elements are irrelevant for color management in photographic instruments, they are of major importance for photographers and other artists. To handle this more general requirement, trichromacy is not sufficient, and a theory of color vision is required that is consistent with the wiring of the brain. This will lead us to *color opponency theory* (Ewald Hering, 1872), which can also account for nonspectral colors such as brown and the inhibitory aspects of blue–yellow, red–green (cyan), and white–black color pairs. Trichromacy and opponency color theories were once thought to be contradictory, but the proper view is that they are complementary, and, indeed, both are necessary for a complete understanding of color vision. In fact, electrophysiology and functional-imaging studies support the idea that trichromacy describes the detection of light by cones in the retina, but that the wiring of nerve cells in the retina and the thalamus is consistent with Hering's opponency theory. This leaves much to understand. Research on the visual system is currently concerned with the mechanism of automatic white balance (color constancy) and the locus in the brain of regions that determine color, form, and position attributes.

15.2 Color Theories

Newton and Young-Helmholtz. Modern theories of color begin with Isaac Newton (1642–1726) (see [Westfall 93]). Imagine the young Newton setting up a lens and a prism in his darkened room at Trinity College, Cambridge, in 1666. A ray of light from the window, defined by an aperture, is focused onto the prism. On the output side of the prism the ray is spread into a strip of light on a screen that glows with the colors of the rainbow. This experiment was not completely new: others had seen colored "fringes" from lenses and prisms and had presented various explanations. René Descartes, for example, had concluded that the colors are produced by "modifications of white light." Newton, on the other hand, was impressed by the fact that the "spectrum" was much longer than it was wide, even though the incident beam had a circular cross-section. To him, the experiment demonstrated the separation of light into its components revealing that "the light of the Sun consists of rays differently refrangible." He followed up this insight by directing light from a narrow (monochromatic) portion of the spectrum into a second prism to demonstrate that a single "color" cannot be further resolved. In addition, he used a

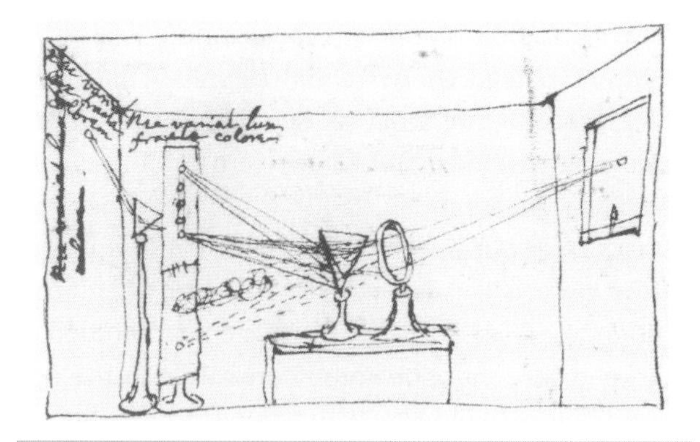

FIGURE 15.2. Newton's sketch of his optics experiment.

lens to refocus the complete spectrum into a spot so that white light was reconstituted. In modern terminology, we say that the components of light are refracted by a prism each according to its wavelength.

Also of note is Newton's construction of a color circle, in which the colors are represented by arcs with lengths proportional to their extent in the spectrum and with red adjacent to violet to connect the ends. This was the first of many geometric constructions designed to permit the estimation of colors obtained by the mixing of light. Newton did not accept the wave theory of light and did not know on what basis the rays were separated by the prism. He advocated a scientific style that relies on experiments and mathematics to demonstrate the truths of nature rather than beginning with hypotheses. He was not, however, immune to preconceived ideas in the interpretation of experiments. Consider, for example, his assertion that the solar spectrum contains seven pure colors. This idea, which was to cause some grief to later investigators, was motivated by his view that the colors in the solar spectrum correspond to the seven notes in the diatonic musical scale (see [Fauvel et al. 88]).

Here is one of many episodes that illustrate Newton's careful experimentation and the relevance of his work to modern technology. In Book One, Part II, of *Opticks*, he reports, "For I could never yet by the mixing of two primary Colours [of light] produce a perfect white. Whether it may be compounded of a mixture of three taken at equal distances in the circumference [of the color circle], I do not know,

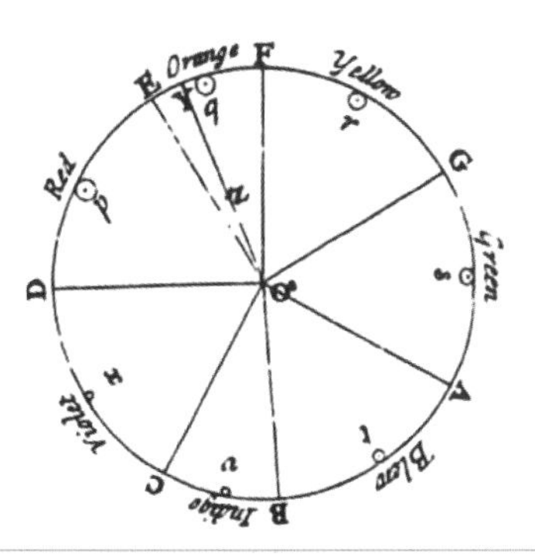

FIGURE 15.3. Newton's color wheel.

but of four or five I do not much question but it may" [Newton 52]. Much of the later history of color management concerns tricolor matching of perceived colors, and in the 21st century there has been much interest in design of white LEDs (light emitting diodes) based on monochromatic LEDs. It has been concluded that at least two complementary light sources with the proper power ratios are necessary for the production of white light, but that a "fuller" (easier and more robust) white can be obtained by adding more colors. Actually, the first commercially available white LED used a blue-emitting diode coated with a yellow-emitting phosphor.

The journey from Newton's initial ideas about light mixing to modern concepts of color management in photography involved contributions from some of the best scientists over a period of 300 years. Briefly, the path proceeds as follows. The polymath Thomas Young (1773–1829) proposed (in 1802) that color vision depends on only three kinds of color-sensitive cells in the retina of the eye that respond to red, green, and violet light. He reasoned that the eye could not have separate receptors for all colors and that three would be sufficient, as shown in Figure 15.4(a). This idea was not confirmed experimentally until 1959. Herman von Helmholtz (1821–1894) extended Young's work on tricolor representation of color by constructing triangles to describe color mixing. He demonstrated that the pigments required for mixing paints are not the same as the colors needed in light mixing, and he introduced the three variables that are still used to characterize a color: hue, saturation, and brightness. In the same time period, James Clerk Maxwell (1831–1879), famous for developing the equations of electromagnetic theory that bear his name, showed that almost all colors can be derived from red, blue, and green light and introduced a two-dimensional color system based on psychophysical measurements. Thus Maxwell was the first investigator to use test subjects to judge how color samples compare with mixtures of primary colors. The relative proportions of three colors were recorded, and since that time these numbers have been known as "tristimulus values."

Light-mixing, such as blue and yellow to produce white, is the consequence of the wiring of the brain, while paint-mixing is simply the physics of light absorption. For example, cyan paint on white paper blocks the reflected red light while yellow paint blocks reflected blue light. If yellow and cyan are both applied, red and blue are both blocked, and we are left with green as shown in Figure 15.4(b). In the subtractive process, the pigments act as filters, and their effects are multiplied.

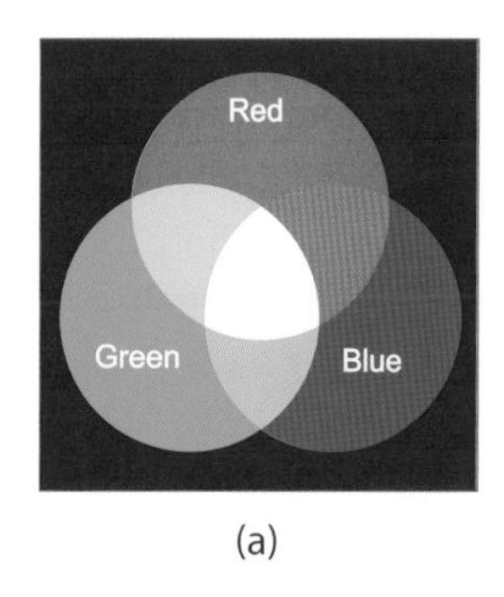

(a)

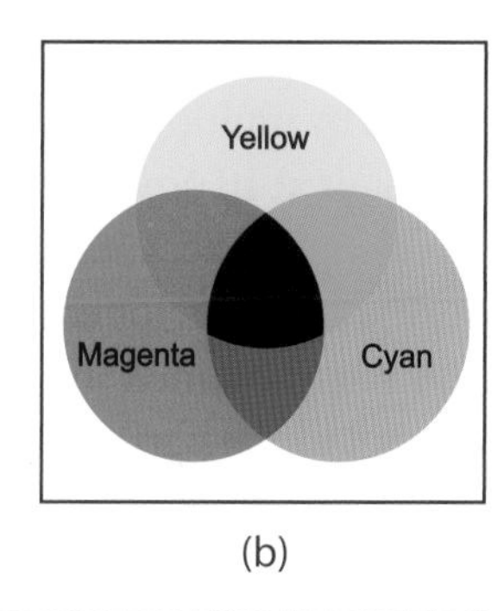

(b)

FIGURE 15.4. (a) Additive color primaries; (b) subtractive color primaries.

15.3 Colorimetry

In this section, we are concerned with specifying color stimuli, acting on the retina of the human eye, on the basis of the Young–Helmholtz color theory. A stimulus is light that has been reflected or transmitted by an object or, perhaps, has originated from a luminous surface. (Each point in a scene contributes a stimulus.) This light is characterized by a distribution of power in the range of wavelengths that affect the cones in the retina. For example, the spectral power distribution of radiation from an object heated to 5000 K is shown in

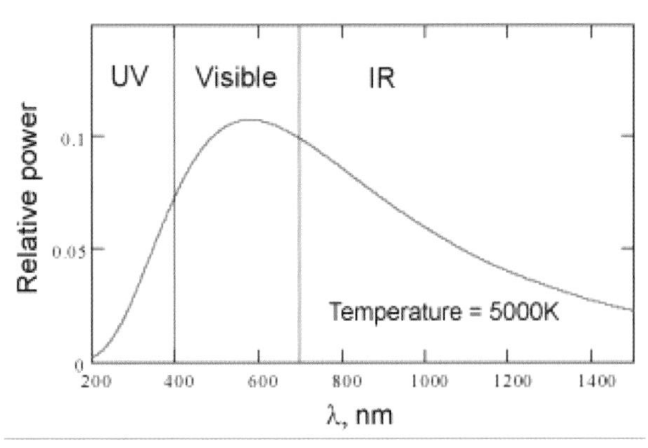

FIGURE 15.5. Spectral power distribution for blackbody radiation at 5000 K.

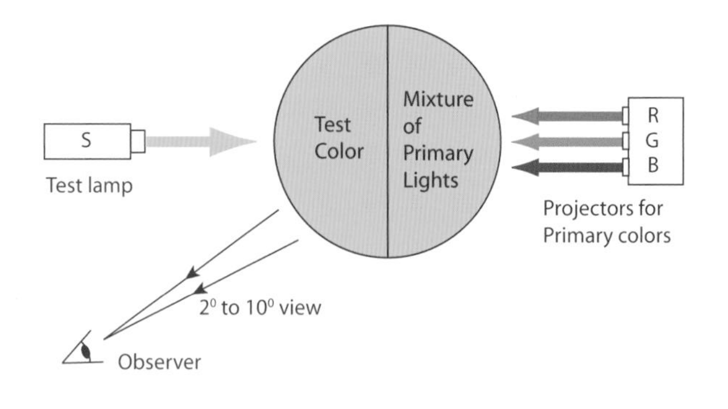

FIGURE 15.6. The classic color-matching experiment.

Figure 15.5. This particular power source approximates the power distribution of the sun at midday and can serve as a "white point" reference.

The retina is not equipped to characterize stimuli by measuring the intensity at each wavelength, but instead responds to the stimulation of the three types of cones. If the sensitivity of a cone is known at each wavelength, the stimulation can be determined by multiplying the power at each wavelength (perhaps at 10 nm intervals) by the sensitivity at that wavelength and adding up the results. The procedure is repeated for each type of cone, and the three numbers obtained are the tristimulus values with respect to the particular set of sensitivity functions. These three numbers specify the stimulus and are correlated with the color that may be perceived; however, it is important to realize that two different stimuli may give rise to the same three numbers. Such sets of stimuli are called *metameric pairs*. This is, in fact, the basis of modern color science since the same perceived color can be obtained without having to reproduce the original stimulus.

The analysis of color stimuli and the determination of when they match is the aim of the science of colorimetry. To understand how this works, imagine an observer sitting in front of a small screen with a test color projected on the left side. The right side of the screen is illuminated by beams of red, green, and blue light (the color primaries) from projectors that are controlled by the observer as shown in Figure 15.6. The observer is instructed to vary the amounts (intensities) of the three colors on the right side until the two sides match in hue and saturation. When a match is obtained, the intensities, or values, of the primaries are noted and another test color is projected. Since the visual field is only 2–10° wide and the surround is dark, monochromatic test colors can be associated with wavelengths. In this way, a table is constructed that contains the three tristimulus values for each wavelength of the color sample (typically about 380 nm to 740 nm). The resulting curves of values versus wavelength are called *color-matching functions* for the particular set of color primaries and the observer.

For example, Figure 15.7 shows the average color matching functions obtained with the CIE RGB primaries defined by red (700 nm), green (546.1 nm) and blue (435.8 nm) for a group of observers with normal vision. At wavelength λ, this result can be described by the equation $S = rR + gG + bB$, where r, g, and b are numbers plotted vertically in the figure and R, G, and B represent the primaries. This equation seems to show that any stimulus is equivalent to a sum of the proper amounts of three primaries, or that any color can be obtained by mixing three primary colors. The figure reveals a region of negative values, however, and this result is found with any set of independent monochromatic primaries. The only way to match the test color in the "negative" region is to add some of the red primary on the test-color (left) side so that the resulting combination can be matched with a combination of G and B. This situation is

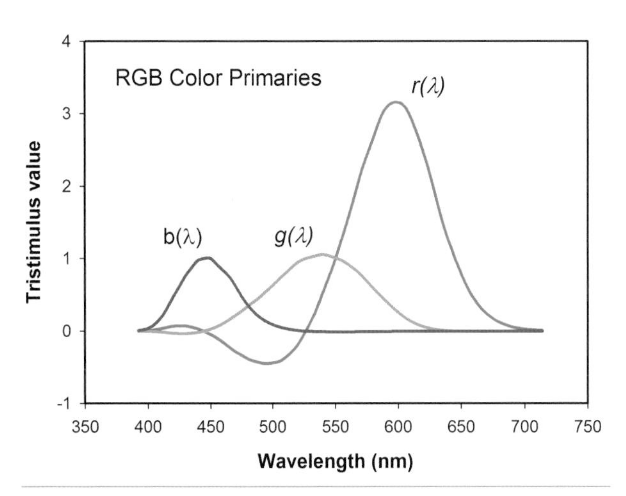

FIGURE 15.7. Matching functions obtained with CIE RGB primary colors.

described by the equation S + rR = gG + bB; by rearranging, we find that S = gG + bB − rR, thus justifying the negative values in the figure.

This brings us to CIE colorimetry. In 1931, the Commission Internationale de l'Éclairage (CIE) met and adopted a particular set of color-matching functions to define the Standard Colormetic Observer (see ["CIE" 09]). At this conference, strong personal opinions and the limitations of technology were accommodated by reasonable compromises. Somewhat different decisions would be made now, but the standard that was adopted served us well into the 21st century. The set of CIE primaries used in Figure 15.7 were adopted, but a system was established that would provide positive tristimulus values. The idea was to represent the characteristics of human observers with normal vision, but the choice was, in fact, somewhat arbitrary since all sets of color-matching functions for a given observer are simply related to each other through mathematical transforms. With any set of monochromatic (single-frequency) primaries, it was found that at least one of the matching functions must be negative at some wavelengths as shown in Figure 15.7, thus a negative tristimulus value would appear. Therefore, three narrow band colors cannot be combined to match all the colors that humans can perceive; and, in fact, cannot really be primaries.

It is clear that true primaries must extend over some range of wavelengths to excite the cones in the human eye selectively. For mathematical convenience, a set of "imaginary" primaries was adopted that exhibit the required properties. That does not mean *imaginary* in the mathematical sense of the word, but, rather, hypothetical illuminants that can have negative power in some wavelength ranges. These hypothetical primaries have the property that they can excite a single type of cone in the human eye even though the spectra ranges of the cones overlap. Therefore, even though they cannot be realized in the physical world, they can provide reference points on the chromaticity diagram and thus serve as the basis of the CIE (1931) color-matching functions that are shown in Figure 15.8.

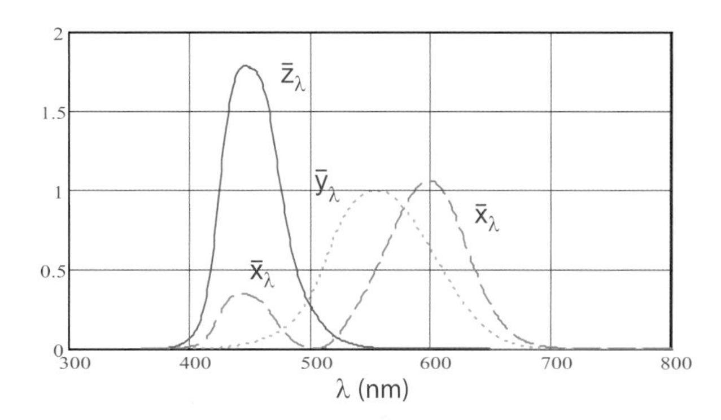

FIGURE 15.8. CIE (1931) color-matching functions for the standard observer.

The matching functions can then be used to compute the CIE tristimulus values X, Y, and Z for any stimulus $S(\lambda)$ by means of the following summations:

$$X = k \sum_{\lambda=380}^{700} S(\lambda)\bar{x}(\lambda),$$

$$Y = k \sum_{\lambda=380}^{700} S(\lambda)\bar{y}(\lambda),$$

$$Z = k \sum_{\lambda=380}^{700} S(\lambda)\bar{z}(\lambda). \qquad (15.1)$$

(When k is chosen so that $Y = 100$ for a perfectly white object, the Y-values are called *per cent luminance values.*) Equation (15.1) is simply a precise shorthand description of the procedure outlined above for determining the effect of a stimulus on each type of cone. Basically, the stimulus values at evenly spaced wavelengths are multiplied by the sensitivities of the cones at the same wavelengths, and the results are added (summed). When the light is reflected from an object, it is appropriate to express the stimulus $S(\lambda)$ as the product of the power distribution of the light source, $I(\lambda)$, and the spectral reflectance of the object, $R(\lambda)$. The following tristimulus value ratios, also known as *chromaticity coordinates*, are defined so that chromaticity can be plotted in two dimensions: $x = X/\Sigma$, $y = Y/\Sigma$, and $z = Z/\Sigma$, where $\Sigma = X + Y + Z$. Since $x + y + z = 1$, all of the coordinates can be represented by plotting y against x as shown in Figure 15.9. From the definitions, all computed chromaticity coordinates are positive, and white is represented by $x = y = z = 1/3$.

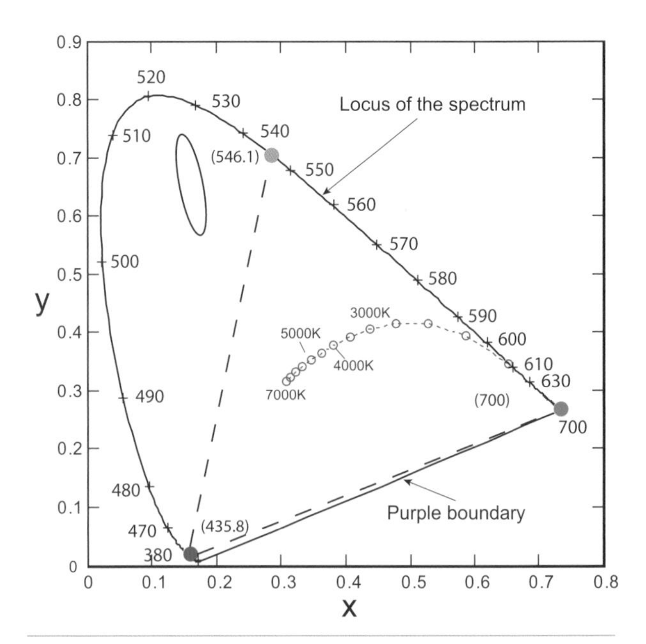

FIGURE 15.9. The CIE chromaticity diagram with the CIE RGB color space.

This diagram should be carefully studied. Light sources that are defined by a single wavelength, i.e., monochromatic light, give stimuli that fall on the locus of the spectrum line. The area defined by the spectrum and the purple boundary contains the coordinates of all stimuli that can be seen by the CIE standard observer. Furthermore, all stimuli that can be obtained by mixing the CIE RGB primaries fall inside the dotted triangle. This triangle represents the gamut of the CIE RGB color space.

Actually, *gamut* refers to the subset of colors limited by a set of primaries including the range of brightness levels from the darkest black to the brightest white. The RGB primaries are "real" since they can be realized in the physical world, but the CIE color space is based on the imaginary primaries located at red ($x = 1$, $y = 0$), green ($x = 0$, $y = 1$) and blue ($x = 0$, $y = 0$). What this means is that, in principal, the "imaginary" red primary, if substituted into Equations (15.1), would give $x = 1, y = 0$, and so on. (For additional discussion, see [Giorgianni and Madden 98]). It is clear from Figure 15.9 that the imaginary primaries must be chosen so that their gamut includes all stimuli that can be seen by the standard observer and that the resulting tristimulus values can be positive. However, the triangle defined by $x = 1, y = 1$, and $z = 1$ necessarily encloses sets of tristimulus values (outside the spectral boundary) that can neither be produced nor observed.

Figure 15.9 also shows white reference points near the center. The blue circles label the chromaticity coordinates for blackbody radiation computed in 500 K steps. The 5000 K (D50) circle was computed using the power distribution shown in Figure 15.5 and the CIE matching functions (Figure 15.8) in Equations (15.1). Coordinates between the 5000K white point and a point on the spectrum curve represent varying degrees of saturation ranging from 100% on the spectrum line to 0% at the white point.

Note. The standard illuminant D50 is constructed to represent natural sunlight (see ["Standard Illuminant" 09], which gives the nitty-gritty of color temperature and standard illuminant calculations). The same chromaticity coordinates can also represent light sources that have spectral-distribution functions that are far from that of sunlight. Hence, 5000 K on a fluorescent label may not mean the same thing as D50. This is not a trivial difference, since absorption, reflectivity, and polarization can depend on wavelength. For example, your jacket may not appear to be the same color when viewed under D50 and 5000 K light.

It is common for the chromaticity diagram to be presented in color with white near the center, then pastels, and finally saturated colors in the spectrum at the edges. This can be beautiful, but it is a fantasy. The CIE color system can represent a stimulus but not a perception. It can be used to specify differences between stimuli but not the appearance of a stimulus. As previously described, the appearance depends on the surrounding colors and the viewing conditions. The same stimulus, in fact, may give the appearance of quite different colors under different conditions.

Two points need to be made before I leave this general discussion of color representation. First, the gamut of accessible colors for a color space depends also on brightness and, therefore, must be represented in three dimensions. The *xy*-plot is, of course, only a projection. The three-dimensional representation shows limitations on our ability to obtain brightness for certain colors. For example, highly saturated colors, other than yellow, are only obtained with low luminance. 3D representations of various color spaces are available online. They are nice to look at, but usually not very useful for photographers. As with the *xy*-plots, there is an irresistible temptation to decorate them with colors. Also, the perception of size or volume of color spaces in 3D can be misleading.

This leads me to the second point. The human eye is not equally sensitive to shifts in the different directions in the chromaticity diagram. If a particular stimulus is selected, i.e., a point in *xy*-space, observers can more accurately match that stimulus with combinations of primaries that vary in a particular direction than with combinations that are perpendicular to that direction. Economic interests have driven a lot of research in this area because of the desire to provide the best representation of color at the lowest cost. Some of the conclusions are summarized in MacAdam ellipses, constructed at various points in the chromaticity diagram, that encompass all the colors that are indistinguishable to the human eye. Roughly speaking, in the region where y is greater than 0.5, the ellipses tend to be long in the direction toward the peak of the spectral horseshoe (520 nm). One MacAdam ellipse is shown in the upper part of Figure 15.9 (see ["MacAdam Ellipse" 09]; experimental results of David MacAdam are plotted on the CIE chromaticity diagram). This emphasizes the need for color spaces that have perceptional uniformity so that the distance between points is correlated with differences in perceived color. The CIE LAB and CIE LUV color spaces are attempts at nonlinear compressions of the CIE XYZ to improve uniformity. These are 3D spaces in which the lightness scale is proportional to the cube root of the luminance as an approximation to our logarithmic response of the human eye to luminance. The CIE LAB color model is now widely used and is available in standard imaging software.

15.4 Color Spaces for Digital Photography

With the almost-universal acceptance of digital photography along with computer processing of images, photographers face the concept of color management daily whether they recognize it or not. We all want our images to have the same appearance on the Camera LCD, the computer screen, in prints, and perhaps in projection as well. This may be taken for granted until obvious and distracting color shifts are noticed, and then we want to fix it. The task is complicated by the fact that each presentation device has a native color space and gamut. The first step in the color pathway is often a digital camera fitted with a sensor that captures the image through three color filters, usually red, green, and blue. Each filter covers about one third of the visible region, and there is some overlap of the regions.

The detailed specifications of sensors are proprietary and often not available to the public. Fortunately, Kodak offers data sheets online for their catalog of sensors. For example, the spectral response of the sensor in the Leica M8 is shown in Figure 15.10. A sensor behind a red filter, for

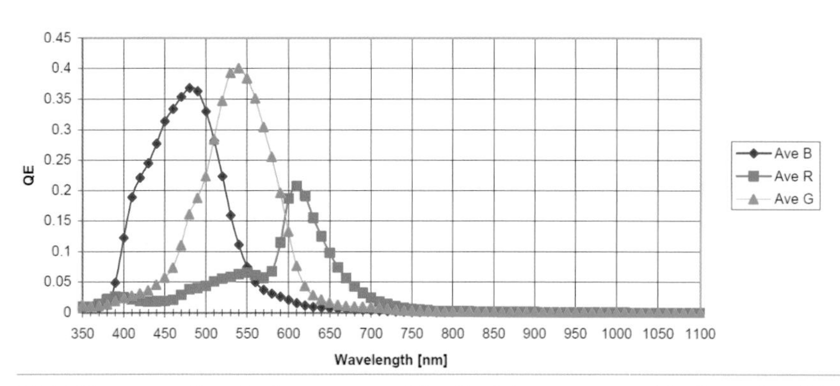

FIGURE 15.10. Spectral response of the Kodak sensor used in the Leica M8. *Used with permission of Eastman Kodak Company.*

example, will accumulate a number of electrons in response to a stimulus, and this number will be digitized and processed off-sensor to somehow represent red. The user does not know for sure even what processing and "error" correction goes into constructing the RAW image.

So what color space should be used in an image-processing program such as Photoshop? The Leica M8 sensor obviously detects the entire spectrum visible to humans and in principle has a gamut greater than the human eye. Why not use a camera-defined color space? That is a possibility, but experts in color management agree that standard color spaces such as sRGB, Adobe RGB, etc., offer advantages. For those who select JPG file format, or have it selected for them, the choice of color space must be made in the camera as it is burned into the file. Full-featured cameras offer the choice of a RAW file format, which leaves the specific camera-based color space in place until the conversion step. At that point the user must assess the advantages of the various alternatives. If the color space is too small, information will be lost and up-conversion cannot recover the information, so it is important to choose the color space with the largest practical gamut initially, realizing that down-conversion to match the gamut of the web or a projector can easily be made later without loss.

Some popular color spaces are shown in Figure 15.11. The sRGB space was designed to make use of the properties of typical monitors and printers in the late 1990s, but, with the advent of large-gamut LCD monitors and printers with eight or more colors of ink, a wider gamut is advisable. Adobe RGB (1998) has been available for several years and is a good alternative to sRGB at least for image processing, but now even Adobe RGB is being challenged. Some professional photographers advocate the use of reference output the media metric (ROMM) or ProPhoto color space. A glance at Figure 15.11 shows that ProPhoto makes use of imaginary primaries and characterizes some stimuli that cannot be displayed or even observed by the human eye.

So why not use the largest possible color space just to be on the safe side? There are two problems. First, the range of colors that must be covered by a certain number of digitized steps is greater even

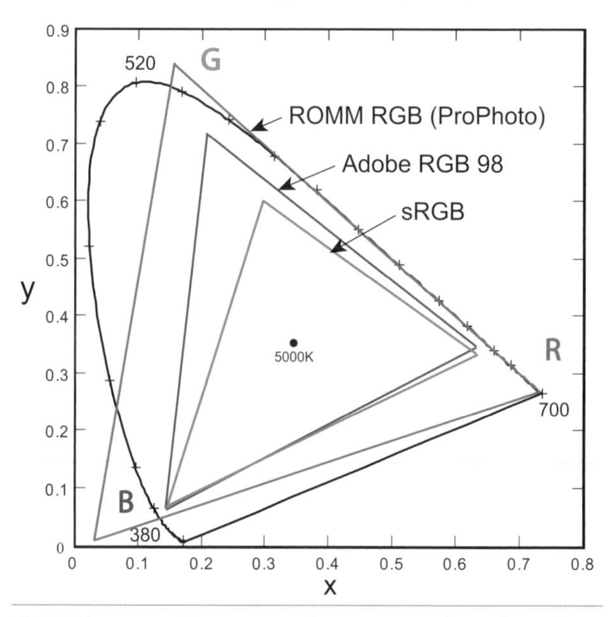

FIGURE 15.11. Common color spaces plotted on the CIE chromaticity diagram.

though part of the range is useless. This can lead to obvious steps in shades of colors when 8-bit digitization is used. With 16-bit processing this is unlikely to present a problem. The second concern is that more information must be stored. The conclusion is that one must decide what is large enough. Professionals and serious amateurs who expect to make large prints with printers that use six or more colors of inks or pigments may wish to select the ProPhoto color space now and to use special proofing techniques to ensure that the additional colors can be displayed or at least handled in a reasonable way. If photographs processed in the wide-gamut color spaces need to be displayed on the web or projected, it will usually be necessary to convert them to sRGB space.

15.5 Color Management in the Tristimulus World

The point of color management is to maintain the appearance of colors in images from the capture (digital camera or scanner) through the display (monitor) and on to the printout (inkjet, laser printer or press profile). This involves an extensive area of convention and technology that is the subject of several monographs. Here is a brief summary of what is involved. Each device is characterized by a profile that specifies at least the gamut (characteristics of the colorants) and the dynamic range (white and black points). The image-processing system in a computer acts as command central. It receives data with input profiles and provides a *profile-connected space* (PCS) selected by the user. In other words, the RGB values from the source are converted to coordinates for the PCS. Another conversion is required to provide RGB values to the monitor according to the monitor profile. (This assumes that the monitor has been calibrated and a profile is available.)

Color reproduction in a monitor depends on the gamut of the monitor, and that is likely to be smaller than the gamut of the color space of the PCS. For example, the monitor may be able to display all colors in the sRGB space but not the complete ProPhoto space. This leads to the little-understood topic of *rendering intents*. That is, if you need to convert to a smaller color space how do you clip colors? Do you throw away the out-of-gamut colors (*relative colorimetric*) or attempt to preserve some relationship between out-of-gamut colors even at the expense of in-gamut colors (*perceptual*)? These are not the only choices, and this is not just an academic exercise since choices must be made at least when printing parameters are selected. Furthermore, a printer will have a different gamut from that of a monitor, and how can we know how the printed image will appear? Fortunately, image-processing programs provide for previewing the printed image on a monitor and detecting the out-of-gamut colors. (For additional information, the reference [Fraser et al. 05] is recommended.)

15.6 Goethe-Hering Theory

The success of trichromacy, or Young-Helmholtz theory, does not mean there has been no controversy. In fact, there is an alternative color theory that has its origins in the work of Johann Wolfgang von Goethe (1749–1832). Goethe is justly famous for his contributions to art, literature, drama, and even science, his best-known work being the dramatic poem *Faust*. Therefore, it is surprising that he wanted to be remembered for his study of the perception of color. From a few observations with a prism borrowed from a friend, in 1790 Goethe became convinced that Newton was wrong about composition of color. This led to many years of experimentation and heated polemic. Goethe misinterpreted his prism experiments and apparently misunderstood Newton's explanations. He favored the Aristotelian theory that colors result from the interplay of dark and light, and there were many other observations that convinced him that Newton had missed important points. Chief among these were the apparent colors of shadows and afterimages of color. Famously, he described staring at a comely girl wearing red in a bar and then seeing a green "after-image" when he looked away. He concluded that there are pairs of colors that stand in opposition to each other and cannot be mixed.

Scientists and many others were repelled by Goethe's attacks on Newton, then dead 100 years, and his promotion of "unscientific" ideas (see [Ribe and Steinle 02; Duck 88; Sepper 88]). He failed to discredit Newton, and he did not provide a satisfactory alternative theory. However, he was on to something that Newton missed. Newton dealt with the physics of light and did not take into account the slippery relationship between subjective colors and "refrangeability" of rays. His writings do indicate that he mused about the problem of how the "Phantasms of Colours" are produced. In fact, so much of what we perceive via vision is determined by physiological factors that many knowledgeable neurologists and even physicists now believe that "Goethe got it right about color." That, of course, does not detract from what Newton got right about light.

In fact, the rehabilitation of Goethe has resulted mainly from the relation between his ideas and the successful Hering theory of color opponency. Ewald Hering (1834–1918) provided another explanation for color-mixing that appeared to be very different from the Young–Helmholtz model. From visual experience he recognized that red, green, yellow, and blue have special standing. In place of the three primary colors (trichromacy), Hering envisioned (in the eye and/or brain) three opponent processes. He proposed that the pairs blue–yellow, red–green, and dark–light control our perception of color. In this way he could explain the strange properties of light-mixing that had been discussed at least since the time of Goethe. First, yellow and blue do not mix, and red and green do not mix. The statement that blue and yellow make green is incorrect for light. The fact that some people believe it probably results from their experience with the mixing of paint or pigment, which is a subtractive (absorption) process. The addition of blue and yellow light, in fact, gives white light if the proper hues and intensities are used. Similarly, the addition of red and green light gives yellow. There are no bluish-yellow or reddish-green colors. Actually, red and cyan give white, so that is probably a better opponent pair.

Photographers will recall the color-balancing tools in programs such as Photoshop, where the sliders range from yellow to blue, cyan to red, and magenta to green. These are antagonistic processes where one color appears at the expense of the other. White and black, of course, depend on contrast, so that black and gray are experienced because they are less bright than the surrounding area. Hering was able to explain the appearance of nonspectral colors by opposing effects of light and dark. For example, brown cannot be matched in the classical color-matching experiment. It only results from our perception of yellow or orange surrounded by brighter light.

The Hering theory had much success, and it won over many psychologists and others. Physical scientists had problems with the theory, however, because it did not seem to have any physical basis. After all, there are three primary colors and trichromacy works in most cases. We now know that Young–Helmholtz theory describes the operation of photoreceptors, and Hering theory is related to the operation of the visual system at later stages in the retina and the brain. Hering was indeed prescient. His insight, or intuition, led him to an amazingly good description of the operation of the opponent visual system 50 years before it could be verified by experiments. This is reminiscent of Young's insight, or guess, that there were three types of color receptors in the eye.

15.7 Monochrome and Monotone Images

There is some confusion about what constitutes a monochrome image. This confusion may arise because terminology in photography differs from that in traditional art. From color theory and photographic usage, it is clear that *monochrome* means black and white. In terms of the RGB color space, every pixel in the image must exhibit three identical values for R, G, and B. For example, the RGB values of a white region might be 255, 255, 255 and those for a black region will be close to 0, 0, 0. At some intermediate value in the gray scale the result might be 50, 50, 50. On the chromaticity diagram (Figure 15.9), only a single point is represented, and that point is near the center of the diagram at a particular white point (depending on the color temperature) denoted by the chromaticity coordinates x and y. This can easily be verified with standard photo-processing programs. In Photoshop, grayscale can be selected in the Mode menu or monochrome can be selected in the Channel Mixer window. The resulting image can be reset to RGB and examined in the Info menu to determine the RGB values for any pixel. This is very different from desaturating two colors or manipulating images in the Channel Mixer, procedures that lead only to different ratios of RGB values in different pixels and another color image, albeit different.

So is it possible to obtain a monochrome image that has a color or hue? This would mean an image in which the RGB values for every pixel have the same ratios of R to G to B values. That is to say, every pixel would be represented by the same point in the chromaticity diagram and would have the same hue and saturation. If the primary color blue was chosen, the values would range from (0, 0, 0) to (0, 0, 255) for 8-bit color and all pixels would have 0, 0 for the R, G values. This would not be very interesting, as there would be no shades of blue, and, in fact, pixels would range from bright saturated blue to very dark saturated blue (black). Furthermore, there would be no white. This kind of image can be viewed (though not printed in color) in the Photoshop Channels Palette by turning off the red and green channels and selecting the color-display option. Of course, some other hue and saturation could be picked for the monochrome image, but white could not appear in the image unless grayscale was chosen. For that reason, there is essentially no interest in single-hue monochrome except perhaps as a faint tint on grayscale images.

In contrast to this clear definition, artists sometimes refer to a painting as monochrome when it is painted with shades of a single color. The problem here is that mixing a pure color (hue) with white changes the position in the chromaticity diagram. Similarly, projecting blue light and white light onto a screen produces a pastel shade that is, in fact, a new color. White light contains equal amounts of R, G, and B, and the blue light only contains B. Therefore, the combined RGB values do not maintain the same ratios from pixel to pixel. This can be expressed in the hue, saturation, brightness (hsb) system by noting that the saturation has changed. Still another example would be an inkjet printer that has only blue ink. The ink dots would either be dense on paper so that saturated blue would appear or sparse so that the white paper and blue dots would simulate other shades. In either case, the shades of blue would be represented in the chromaticity diagram on a line connecting the blue hue with the white point. What I am describing is actually a *monotone* image, not a *monochrome* image. A monotone image can be produced in Photoshop by grayscale conversion followed by selection of Duotone. The monochrome image can then be converted to monotone by selecting one ink only.

The kinds of images discussed here are shown in Figure 15.12. If the RGB values of these images are examined (for example, in Photoshop) the values for the color image will be found to range from (0, 0, 0) at the darkest point to

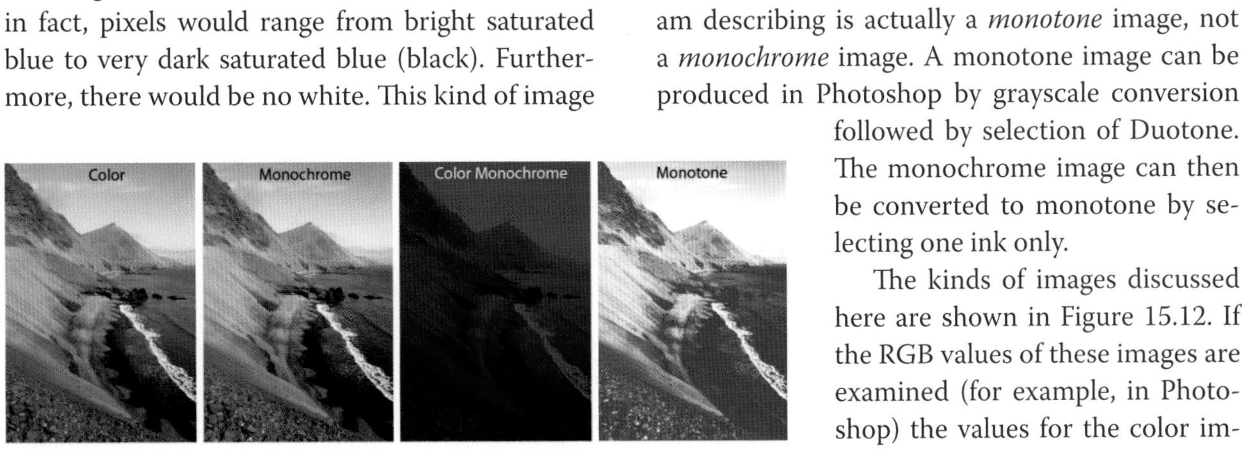

FIGURE 15.12. An image presented in color, monochrome, color monochrome, and monotone.

(255, 255, 255) at the brightest point with a wide range of values and ratios of values in between. The monochrome image also ranges between (0, 0, 0) and (255, 255, 255) but every set contains three identical numbers. The color monochrome should have RGB values that range from bright blue (0, 0, 255) to near black (0, 0, 0) with R and G values always zero. Obviously, there is no white in this image. In an interesting contrast, the monotone image ranges from (255, 255, 255) to (0, 0, 255) with the R and G values always equal. Therefore, white is present but there is no black. In conclusion, for all practical purposes, monochrome means black and white. Monotone (one-ink) and duotone (two-ink) images may be of interest for special purposes such as advertising illustrations. See, for example, the monotone package from Twisting Pixels.

15.8 Color Constancy and the Land Theory

Edwin H. Land, the founder of Polaroid Corporation, reported a set of experiments in 1958 that amazed even the experts in color theory [Land 83]. In a research project that was associated with the development of Polaroid color film, Land and his co-workers were repeating the classic color-photography experiment of Maxwell. This required them to take three images of a scene on black-and-white film using red, green, and blue filters. From these images, slides were made that could be projected in register on a screen. The image acquired with a red filter was projected through a red filter, the blue one through a blue filter, and the green one through a green filter. This, of course, worked, and a brilliant color image was obtained. The principles are known and form the basis of commercial three-color projectors, but an unexpected thing happened at closing time. The blue projector was shut down and the green filter was removed from the green projector leaving only a black-and-white image (from the slide taken with a green filter) and the red image. Before shutting down the red projector, Land noticed that the screen still showed a full-color image instead of the expected pink image. At the time, he thought it might have resulted from some kind of adaptation process in the eye, but later he realized that something remarkable was happening. This prompted him to undertake decades of research on the perception of color and ways to model the operation of the visual cortex.

In later experiments, a Mondrian-like patchwork of colored papers was used to test color perception under a variety of illumination conditions. The major finding was that a given color patch, say the green one, maintains its green appearance in the context of the surrounding patches even when the illuminating light sources are changed so that the stimulus from that patch equals the stimulus that previously had been measured for a blue or a red patch. It turns out that color-contrast across borders is all-important for color vision. In his 1983 follow-up, Land emphasized his conclusion, "whereas in color-mixing theory the wavelengths of the stimuli and the energy content at each wavelength are significant in determining the sense of color ... in images neither the wavelength of the stimulus nor the energy at each wavelength determines the color. This departure from what we expect on the basis of colorimetry is not a small effect, but is complete." In the Retinex theory proposed by Land, there are three mechanisms at work to determine the lightness of areas at three different wavelengths. The important fact in color vision is that an image is simply an array of areas with different lightnesses relative to each other. The computational challenge in Retinex theory is to show how color can be predicted from the lightness of the various areas without any information about the illuminant.

15.9 Color Opponent Cells in the Retina and Brain

Various aspects of our visual experience are captured by the color theories described above, but we need to discover the underlying physical basis for these effects. The lucky "guesses" by Young and Hering brought us the

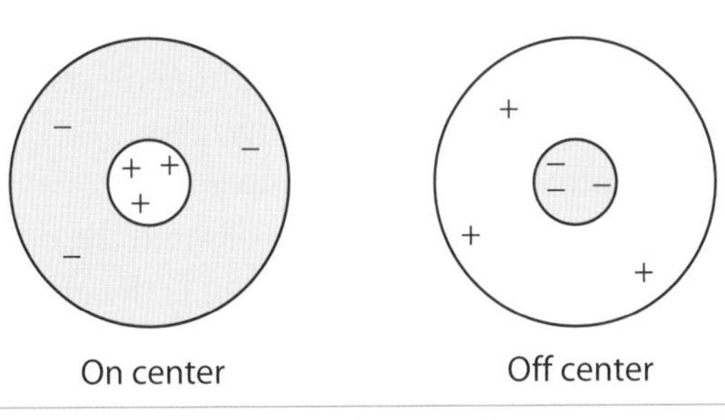

FIGURE 15.13. Luminance (ganglion) cells in the retina.

immensely important ideas of three color receptors and opposing colors, respectively, long before there was physical evidence for mechanisms behind them. But in the last few decades, brain research has advanced to the point that we now have physical models for the visual system that can reconcile trichromatism and color opponency. It is clear that our visual system operates through the transfer and the processing of information about images. The story proceeds as follows. The cones, S, M, and L (for short, medium, and long wavelength, respectively), are connected through bipolar cells to ganglion cells in such a way that the ganglion cells respond to sums, differences, and perhaps other combinations of cone signals. In this way, arrays of ganglion cells with different types of receptor fields are formed.

The luminance cell, which responds to the sum S + M + L, is arranged with a center that responds to a light signal and a surround field that suppresses the response (Figure 15.13), or the reverse where

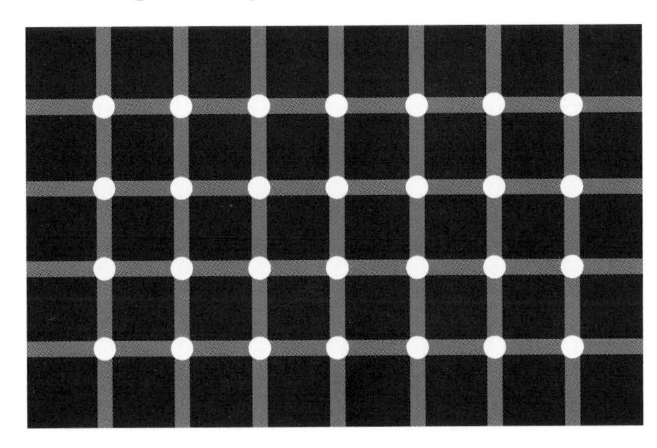

FIGURE 15.14. Elke Lingelbach's version of the Hermann grid illusion (1870).

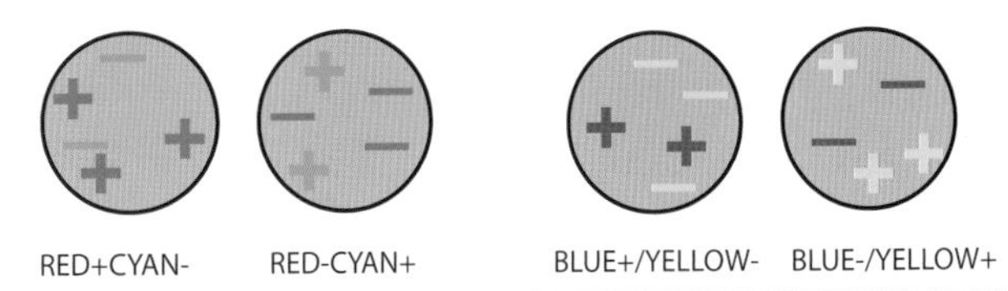

FIGURE 15.15. Single-opponent cells.

the on-center signal suppresses the signal and the surround field responds. In the classic experiment of Hubel and Wiesel, a spot of light was found to increase ("on cell") or decrease ("off cell") the firing of a nerve fiber. In each case when the size of the spot was increased, the response was reduced by the action of the surround ("off center") cells as shown below. The size of the center field depended on the position of the luminance cell on the retina, but was typically about 3° of arc.

The opponent aspect here is spatial, and black and white are not opposite colors. We can, of course, experience any shade of gray. The striking visual effect exhibited by the grid in Figure 15.14 has been used as an illustration of the suppression of the center by the surround field. It turns out that the situation is not so simple, and this explanation is controversial.

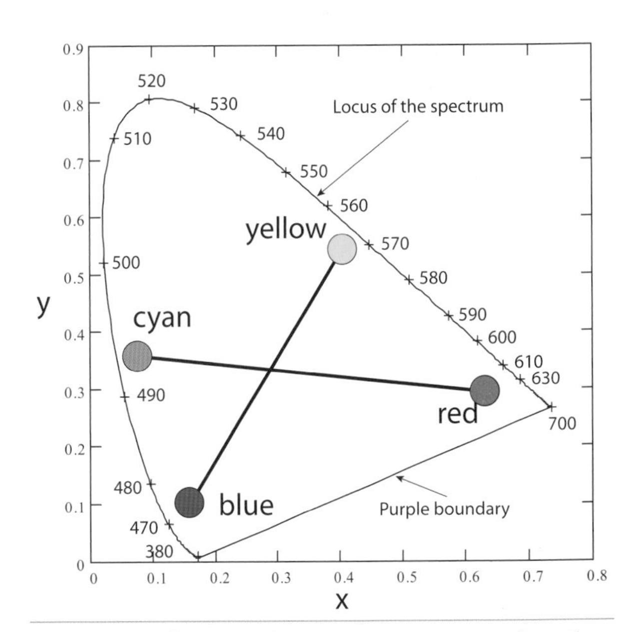

FIGURE 15.16. The color-opponent axes against the CIE diagram.

Things get even more interesting with color. Here the wiring of the nerve field transforms from RGB cone signals to opposing signals on the axes shown in Figure 15.16. For example, we find cells that can be excited with red light (L) and suppressed with green (or cyan) light (M) or vice versa. These simple R+C- and R-C+ opponent cells have co-existing L and M cones, and we find similar arrangements of L + M (yellow) versus S (blue). In Figure 15.15, a plus sign indicates excitation and a minus sign means suppression. A cell either transmits red or cyan but not both, and, similarly, a cell transmits either blue or yellow, but not both. The relation of the opponent colors to the CIE chromaticity diagram is shown in Figure 15.16.

Transmission of information from the retina is in opponency form, and later processing in the visual cortex makes use of more sophisticated opponency cells. For example, a pair of double-opponency cells are shown in Figure 15.17. The red–cyan double-opponency cells require excitation in the center cell by red cones and excitation in the off-center (surround) by cyan cones to be able to fire a nerve fiber.

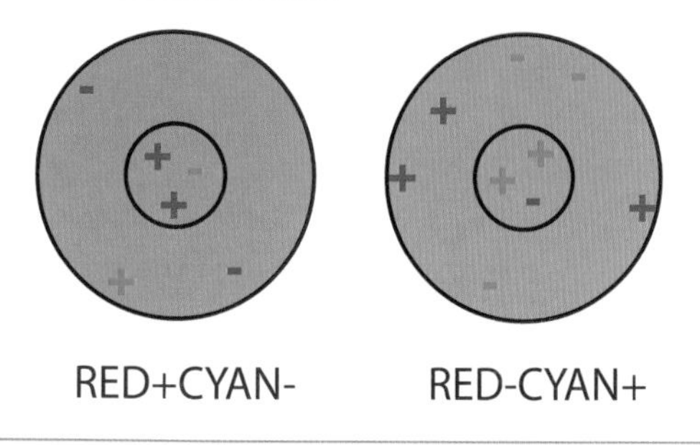

FIGURE 15.17. Red–cyan double-opponency cells.

The opponency structure reminds one of the Hering ideas about opponency, but the variety of simple- and double-opponency cells is much richer than that suggested by Hering's theory. The question naturally arises about why nature arranged things this way. One suggestion is that the luminance system evolved first and that the color-opponency system was added on top of it. Furthermore, the separation of the M and L cones from a single common ancestor cell was a fairly recent event in the evolution of the color system that is only found in old-world primates. This resulted in additional types of opponent cells and programming for information transfer.

Undoubtedly, the major advantage of opponent coding by the luminance cells and the double opponency cell in the color system is the ability to detect "edges." Any abrupt spatial change in color or luminance will cause an imbalance in the center and off-center cells and will lead to increased firing of nerve fibers. Similarly, areas of uniform luminance or color will tend to balance excitation and suppression and will reduce the rate of nerve-firing. In this way colors can be established at the boundaries of spatial regions, and the interior regions can be filled in with relatively little information. Image compression in modern computers is based on similar ideas. If a color is constant for some distance in an image, there is no reason to record its value at every point. For example, blue may extend for 1000 pixels in an image, and in each pixel blue has the value 255. Instead of recording 1000 values of 255, it can just be noted that 255 occurs 1000 times.

The Nobel prize was awarded in 1981 to David Hubel and Torsten Wiesel for their studies of information-processing in the visual system, and research continues to reveal the features of this amazing system (see [Hubel 95; Livingstone 02]). However this work is just at its beginning, because the really hard part concerns high-level processing in the visual cortex and beyond. Even our understanding of the parts that have been extensively studied is incomplete. For example, the ratio of L to M cones in the average eye is about 2 : 1, but this ratio varies by a factor of 4 for different individuals apparently without affecting color vision. The S cones only make up about 8% of the total, and there are essentially no S cones in the fovea. Furthermore, the distribution of L and M cones in the retina is not uniform. One wonders how the receptor fields for the ganglion cells can be constituted.

The brain appears to be performing magic when it converts streams of encoded information from two eyes into an illusion of a three-dimensional full-color world. In fact, the work recently reviewed by Werblin and Roska identifies two-dimensional arrays of 12 types of ganglion cells in the retina that transmit partial representations of the visual field as "movies" to the brain [Werblin and Roska 07]. One image stream contains edges information, another shadows, another motion and direction, etc. This is done in real time with a large amount of feedback at each stage. Thus, our reality depends on the brain's interpretation of an array of continuously updated partial visual representations.

Vision is the primary way we obtain an understanding of the world. It also provides great enjoyment and sometimes awe and wonder. The introduction to the visual system presented here reveals practical aspects of color management, and it also gives a peek at the array of special effects we experience because of the particular architecture of the visual system. By this, I mean color constancy, color groupings that are striking, optical illusions that result from opponent cells, etc. In Chapter 18 I consider perhaps unanswerable questions concerning human creation and appreciation of art. Can science help us understand why humans are attracted to art and fiction in general? Artists are perhaps acting as neurologists as they experiment with the framing of scenes, use of color, representation of light, etc., to determine what is pleasing to the human brain.

Further Reading

S. J. Williamson and H. Z. Cummins. *Light and Color in Nature*. New York: Wiley, 1983.

S. Zeki. *A Vision of the Brain*. Oxford: Blackwell Science Ltd., 1993.

S. Zeki. *Inner Vision: An Exploration of Art and the Brain*. New York: Oxford University Press, 1999.

P. Lennie. "Color vision: Putting it Together." *Curr. Biol.* 10 (2000), R589–R591.

P. Gouras. "Color Vision." Webvision: The Organization of the Retina and Visual System. Available at http://webvision.med.utah.edu/color.html, 2009.

H. Kolb, E. Fernandez, and R. Nelson. "The Perception of Color." Webvision: The Organization of the Retina and Visual System. Available at http://webvision.med.utah.edu/KallColor.html, 2009.

B. MacEvoy. "Color Temperature." Available at http://www.handrpint.com/HP/WCL/color12.html, 2009.

S. McHugh. "Color Management: Color Space Conversion." Available at http://www.cambridgeincolour.com/tutorials/color-space-conversation.htm, 2009.

CHAPTER 16

Image Capture and Processing

Teams of Bell Labs scientists, such as Shockley, Brattain, Bardeen, and many others met the challenge—and invented the information age. They stood on the shoulders of the great inventors of the 19th century to produce the greatest invention of our time: the transistor.

—HTTP://WWW.IDEAFINDER.COM/HISTORY/INVENTIONS/TRANSISTOR.HTM

Humankind eventually would have solved the matter, but I had the fortunate experience of being the first person with the right idea and the right resources at the right time in history.

—JACK KILBY

16.1 Introduction

Photography, or "writing with light," is all about capturing images. Until the 21st century that meant using photographic film and developing it to produce negatives or slides. Now only old-timers remember or care much about film, and it appears to be on the way out. I think film is mainly of historical interest, but it is instructive to compare its characteristics with those of digital sensors. Also, the principles for capturing and reconstructing color images still apply.

The important difference between photographic films and digital sensors comes down to how the image is sampled and recorded so that it can be reproduced with fidelity. The image of concern here is a pattern of light projected by a lens onto a sheet of film or a digital sensor. Recall from Chapter 2 that light consists of packets of energy called photons. To be more precise, the energy of a photon is given by $E = h\nu$ where ν is the frequency of the light and h is a fundamental constant of nature

($h = 6.626 \times 10^{-34}$ J · s). It is an interaction of photons with matter, usually electrons, that permits light to be detected and its intensity to be measured. This sounds abstract, but it is the essential part of understanding how photographic images are recorded.

In order to understand what photons do in film photography, it is necessary to look at the photochemistry and photophysics at the atomic level. This is indeed interesting, but I have relegated it to Appendix F so that we can proceed immediately to the operation of digital sensors, including sensitivity, resolution, and image processing. There are still some areas where photographic film excels, but with the rapid advance of digital technology, it is clear that the use of film will continue to decline. In the 21st century I expect that film will be viewed as no more viable for photography than vinyl records are for audio reproduction.

16.2 Capturing Photons with Image Sensors

Sensors have all but replaced film for image capture, but this is just the tip of a mountain of advances in science and technology. Sensors and their use would be impossible without the microelectronics industry and the plethora of new products, especially microprocessors. The information revolution, based on microelectronics, expands exponentially, sweeping away last year's marvels, and presents seemingly unlimited features and opportunities. By 2008 we had tiny mobile telephones that could surf the Internet, take and transmit photographs and movies, display geographic coordinates and maps, and on and on. Our cameras are now part of the information explosion just as computers are, and we can expect that they will become obsolete every couple of years. The rate of change will eventually slow down as it has for other mature technologies such as television and automobiles, but the transition period is far from over for cameras. The history of microelectronics and the path to the digital sensors now available to photographers is a fascinating story. It is summarized in Appendix G.

16.3 CCD and CMOS Image Sensors

Both CCD and CMOS sensors were being used in digital cameras by 2008, and the difference was not obvious to most users (see [Turchetta et al. 09; Spring et al. 09]). Their names indicate their construction, which is not of direct interest here. This chapter will focus on the way the sensors work and their relative advantages, at least at this stage of their development. In both CCD and CMOS sensors the pixel array can be thought of as an array of small buckets designed to catch photons. These buckets are usually square in cross-section, but may have a variety of shapes. For example, Fujifilm uses octagonal pixels in their Super CCD sensors and Nikon tried rectangular pixels in their D1X DSLR (see Figure 16.1). In DSLR cameras, the pixel pitch (edge-dimension) is typically 5 to 8 μm, while in point-and-shoot cameras it may be less than 2 μm. A photon detector occupies most of the surface area of each pixel in a CCD sensor, while in CMOS sensors the detector must share the pixel with other circuit elements. Therefore, there is an advantage to the CCD design that is more significant for very small pixels. The advantage is shrinking with time, however, because circuit elements are becoming smaller, there is a more efficient use of space in a pixel, and pixel-covering micro-lenses are being used to focus incident light onto the detector.

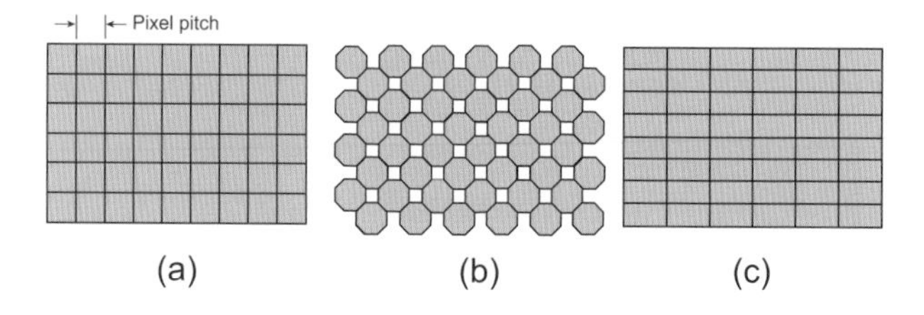

FIGURE 16.1. Pixel arrays: (a) square, (b) octagonal, and (c) rectangular.

In a photon detector, an absorbed photon produces an electron-hole pair and thereby makes a mobile electron available for storage in a capacitor. The *quantum efficiency* of the detector is defined as μ = (number of electrons/sec.)/(number of photons/sec.).

This number depends on the frequency (color) of the light, but for modern sensors the number can be greater than 0.9, meaning that the sensor has more than 90% efficiency. By this measure, the efficiency of film is very low, since only about 2% of incident photons are absorbed by silver halide crystals to produce a latent image. As the exposure increases, the number of electrons in the pixel capacitor increases toward a maximum capacity somewhere between 5,000 and 100,000 electrons, known as the full-well capacity (FWC). The FWC, which determines the maximum signal and the dynamic range, depends on the physical size of the pixel. In CCDs, the charges are shifted cell by cell to external amplifiers. In contrast to this, the electrons accumulated during the integration period in a pixel of a CMOS sensor are stored in the silicon-potential well, and at the end of the collection period are converted to a voltage. Each pixel has a reset transistor, an amplifier, and row and column connections that permit the pixel to be read directly. The construction details and the performance specifications for modern sensors are proprietary and are not readily available from the vendors, but there is no shortage of individuals eager to test the final products.

16.4 Encoding of Color

As described above, the pixels in an image sensor accumulate an electrical charge that is proportional to the intensity of incident light in a certain frequency range. This provides information about luminance but not about color. In order to encode color, we require information about the intensity of light in at least three different frequency ranges. Ideally, each pixel would contain three independent detectors that would mimic the sensitivity of the three cones in the human eye, but this is not yet possible, and we must consider various compromises. One possibility is to use a trichroic prism assembly to reflect red, blue, and green light from the image onto three different image sensors as shown in Figure 16.2. Information from the three sensors can be brought into register and combined in a computer memory to simulate an array of three-color pixels. This scheme, known as 3CCD, works well and is used in some video cameras and camcorders, but it requires three times as many elements for the image detector and is expensive.

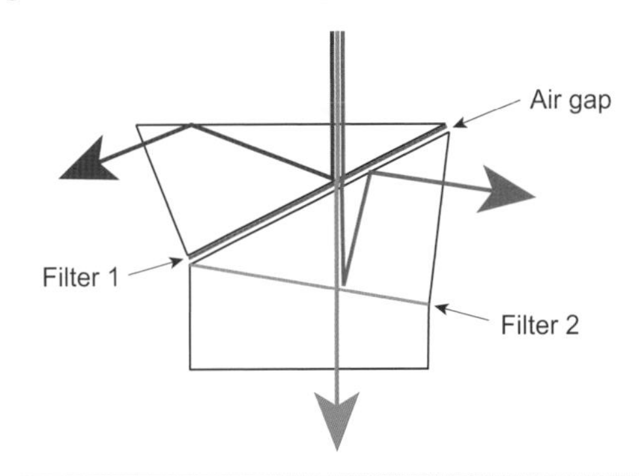

FIGURE 16.2. Phillips-type trichroic prism assembly (see Wikipedia).

Another possibility is to use the fact that silicon absorbs different amounts of red, blue, and green light so that different frequency distributions can be found at different depths in a chip. The silicon acts like a set of filters absorbing blue more than green and green more than red as illustrated in Figure 16.3. This means that detectors at different levels in the chip can obtain enough color information to permit accurate three-color pixels to be simulated. The silicon filtering effect is exaggerated here since all three detectors receive some R, G, and B photons, and only the relative amounts are different. This layout of detectors is, in fact, the basis of Foveon X3 technology that is now being offered in cameras made by Sigma Corporation. Each pixel is capturing most of the incident light by means of three detectors, and the color processing uses only information from that pixel. Therefore, resolution is, in principal, enhanced compared to arrangements where some color information must be obtained from neighboring pixels.

Most digital cameras encode color by using a color filter array (CFA). In this scheme each pixel is covered with a filter, usually red, blue, or green, and information must be combined from a set of neighboring pixels to assign R, G, and B values to a pixel. In other words, after an exposure each

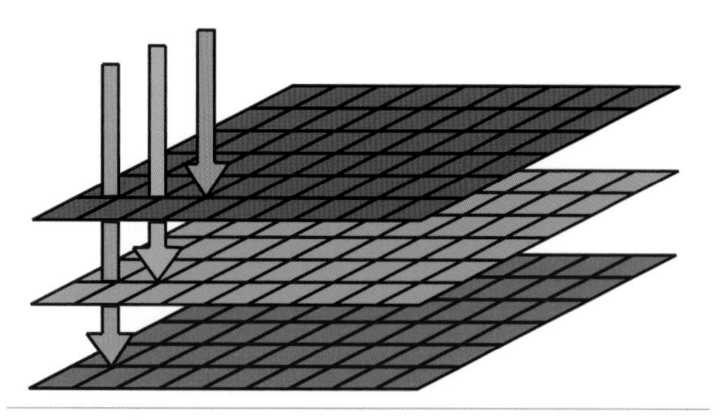

FIGURE 16.3. Illustration of red, green, and blue detectors in each X3 pixel.

pixel has a value (number) that corresponds to the intensity of light transmitted by one of the color filters. The array of numbers for all the pixels must then be processed to obtain a set of three numbers for each pixel to specify its full color. This computation has been called *demosaicing*. In the simplest algorithm, a pixel is assigned color values from neighboring pixels to complete the set of three colors. For example, a red (R) pixel will be assigned B and G values from its neighbors. This gives a fast but poor representation of color. The next-best scheme involves computation of the average color-values of neighboring pixels. Thus, an R pixel gets a B value that is the average of two or four neighboring blue pixels. Still more sophisticated algorithms, including cubic-spline averaging, are available in the integrated circuit inside the camera to determine appropriately averaged values of three colors for each pixel in the sensor.

Filter arrays turn each pixel into a color detector, reminiscent of the operation of the retina in the human eye, especially the fovea (see Chapter 14). A major difference is that CFAs contain periodic patterns of filters and, therefore, sample images at regular intervals, while the cones in the retina have a random distribution. Random distributions are also found for grain in film and the dyed starch grains in the Autochrome process. Unfortunately, periodic sampling can produce undesirable artifacts such as Moiré fringes. Ways to avoid this problem are discussed below. The use of randomly distributed pixels in an image sensor appears to be an impractical solution because of problems with pixel readout and image processing. The human brain requires years to learn how to process the visual signals from randomly distributed cones and rods in the retina. Of course, this is not a fair comparison because the brain must also deal with binocular vision, motion detection, coordination of motion, etc.

By far the most popular filter array for digital cameras is the one invented by Bryce E. Bayer at Eastman Kodak in 1975 (see Figure 16.4(a)). The Bayer pattern is 50% green, which is reasonable, since the human eye has its maximum sensitivity and resolving power in the green region of the spectrum. In this scheme, the green elements determine the luminance while the red and blue elements provide chrominance information.

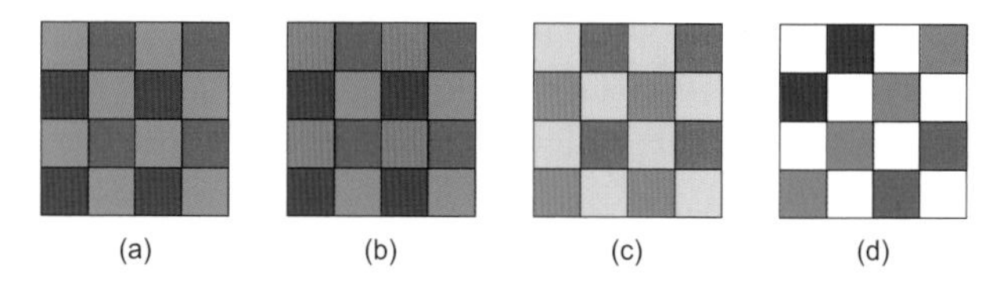

FIGURE 16.4. Color filter arrays: (a) Bayer pattern—RGBG, (b) Sony RGB+emerald pattern, (c) Bayer with subtractive primaries, and (d) Kodak RGB+white pattern.

In the second pattern (Figure 16.4(b)) half of the green filters are replaced with "emerald" filters. This pattern, introduced by SONY, in principal permits a more accurate representation of color than can be obtained with only three filters. As explained in Chapter 15, any choice of three color primaries cannot exactly match all the colors that humans perceive. In fact, with some choices of reference colors, the matching of a test color in the blue–green spectral region requires a negative amount of red light. This simply means that a mixture of the blue and green reference beams can match the sample in a given frequency region only if red light is first added to the sample! This longwinded explanation is meant to justify the introduction of a fourth primary to improve color matching over the entire visible region.

The third pattern (Figure 16.4(c)) is a Bayer array with R, G, and B replaced with the subtractive primaries, magenta (M), yellow (Y), and cyan (C).

Kodak used this one in the DCS60X camera. The justification for the MYCY array is that these filters transmit twice as much light as the RGBG array. The R, G, and B filters each transmit a maximum of 1/3 of the total intensity. In order to understand this effect, think of M = B + R, Y = R + G, and C = B + G. The improved sensitivity may be offset by less-effective demosaicing algorithms and poor color rendition. Some video cameras use a modified MYCY array in which half of the yellow pixels are replaced with green in an attempt to improve the color quality.

The fourth pattern (Figure 16.4(d)) is one of a group of filter arrays that include clear filters to make panchromatic pixels. The panchromatic pixels are, of course, more sensitive, and they provide the luminance information. Kodak is pushing this approach now as a way to improve sensor performance. Various patterns, some including 50% panchromatic pixels, require different levels of image processing and may be appropriate for different segments of the camera market. The main advantage of sensors with panchromatic pixels appears to be sensitivity.

Sensor development will continue to be an active area for years to come. In addition to the steady increase in pixel count, there are reports almost monthly of new schemes to improve sensitivity and dynamic range. For example, the improvement of micro-lens design and placement is continuing, and Fujifilm has attempted to extend the dynamic range of their sensors by including a second detector in each pixel with a different sensitivity.

16.5 How Film/Sensor Speed (ISO) Is Defined

Film "speed" refers to the film's sensitivity to light and how quickly a good exposure can be obtained. Speed and ISO are used interchangeably, but ISO refers to a set of standards published by the International Organization for Standardization. ISO and the older ASA standards are essentially the same for film. Unfortunately, the whole area of film and sensor standards is complicated and tedious. However, two points merit our interest. First, on what measurements are the standards based, and, second, how do film and sensors differ in their response to photons? The quantitative characterization of film dates from the 19th century when Ferdinand Hurter and Vero Charles Driffield started studying plate density as a function of exposure. A characteristic curve (*H* and *D* curve) for monochromatic film is shown in Figure 16.5. The development time determines the steepness of this curve, so the standard is based on film developed to the point that the optical density (OD) changes by 0.8 units from the 0.1 line when the logarithm of the exposure increases by 1.3 units. (The illuminance is measured in units of *lux* (lx), defined in Appendix H.) This shape can be obtained regardless of film speed, but the numbers on the abscissa will, of course, depend on the film. Any film will have some minimum density that arises from the cellulose base and emulsion fog. Therefore, to avoid the density not associated with exposure, the characterization of the film begins at 0.1 OD units above the intrinsic density; and film speed is based on the exposure at the point *m*, denoted by H_m, that is required to reach that level.

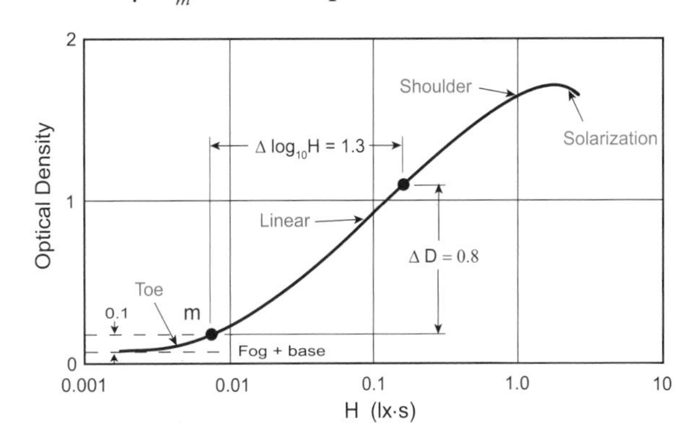

FIGURE 16.5. The characteristic (*H* and *D*) curve for monochromatic film. Optical density versus exposure *H* in lux seconds.

For monochromatic film, the ISO speed is defined as $S = 0.8/H_m$, and a good exposure for an average outdoor scene would correspond to roughly $H_0 = 9.4H_m$. The procedure is more complicated for color-negative film since there are three layers, and color-reversal film requires a different reference point. In all cases, however, the speed is based on the exposure required to have a defined effect on the density of the developed emulsion. After the film speed has been established according to the ISO standards, photographers are able to set their cameras to obtain consistent exposures.

The idea behind the concept of speed ratings is that total exposure determines the density, and the exposure is always the product of illuminance and time. This is the assumption of *reciprocity*, namely, that there is an inverse relationship between the intensity and duration of light required to have a defined effect on film, and photons all count the same regardless of how fast they accumulate. In real life, the reciprocity law fails for exposures less than about 1/1000 seconds and greater than a few seconds. At very short exposures, many photons arrive at the same time, and they are not used efficiently. With very long exposures, there is low-intensity reciprocity failure. In this limit, photons are apparently lost or somehow forgotten.

The *H* and *D* curve reveals other important characteristics of film. In the toe region, starting from zero exposure, we see that many photons are required before there is a measurable change in the density. Next is the linear region, the most useful part for creating images. The slope in this region, defined as $\gamma = \Delta OD/\Delta(\log H)$, determines the contrast of the image. For monochromatic negatives with normal development, gamma is about 0.7. At still-higher exposures, we encounter the shoulder and finally the solarization region, where additional exposure actually decreases the density.

Now consider the characteristics of sensors, either CCD or CMOS. A photo detector (a photodiode) in each pixel captures photons efficiently and activates approximately one electron for each photon. The exact number of electrons per photon is equal to the *quantum efficiency* of the detector. The activated, or free, electrons accumulate during the integration period in a potential well and constitute an electrical charge that can be converted into a voltage. Think of the electrons as filling a well so that the number of electrons at any time is proportional to the exposure. To a very good approximation this is a linear process, and for all practical purposes the reciprocity law is satisfied. Of course, at extremely short and long times there will be variations, but that need not concern us. The characteristic curve of the number of electrons versus the exposure for sensors is shown in Figure 16.6. In this plot the plateau on the right side represents the FWC, or saturation number of electrons.

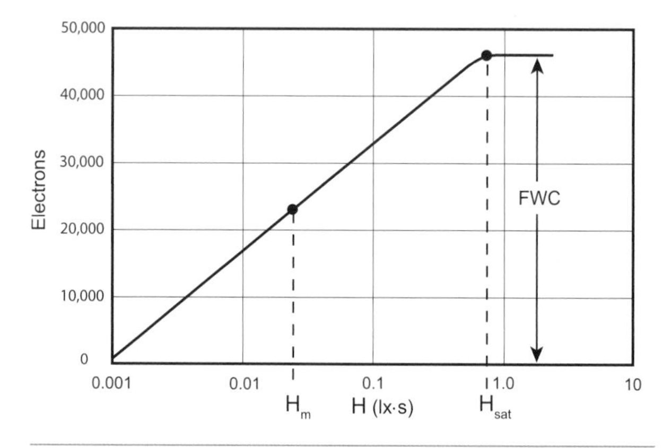

FIGURE 16.6. Characteristic curve for a sensor: electrons accumulated in a pixel well versus exposure *H* in lux seconds.

The speed of a sensor is defined by the standard output sensitivity (SOS), roughly equivalent to the ISO speed, or by the recommended exposure index (REI). SOS and REI are standards published by the Camera and Imaging Products Association (CIPA) (see [CIPA 09]). The SOS speed *S* is based on the exposure H_m that is required to obtain 0.461 of the maximum digital output under prescribed illuminance conditions, and the speed is defined as $S = 10/H_m$. (The factor 0.461 is deemed to give medium output, which corresponds to the relative output of an 18% reflective standard.) Now, suppose that the signal from the pixel is amplified by a factor of 2 before it is digitized so that a full-scale reading (maximum brightness) can be obtained with only half the FWC. As a result, the measured value of H_m will be a factor of 2 smaller and the SOS (or ISO) number will double. The REI is similar, except that H_m is replaced by the manufacturer's recommended average exposure in the focal plane. Still other definitions have been proposed, for example, $S_{sat} = 78/H_{sat}$, where H_{sat} is the exposure that causes clipping of the signal. The minimum value of S_{sat} is determined by the sensor, but higher values can be obtained with amplification before the analog-to-digital (A/D) converter. There are also the noise-based sensor speeds with reference exposures that produce output signals 10 or 40 times the noise in the sensor.

The point of these standards is to establish a number that will yield images similar to those obtained with film at the same ISO setting. So what do camera manufacturers actually report for the SOS/ISO number? It is not always clear. It would be reasonable to use the 10 : 1 noise result as the upper limit and the 40 : 1 noise result or the H_m based speed as the standard value. It is clear, however, that high SOS values are associated with lower-quality images, and the acceptable noise is a subjective judgment.

16.6 How the Dynamic Range Is Determined

The *dynamic range* is the ratio of the maximum to the minimum signal that can be measured. In photography, *signal* refers to the light-sensitivity range of films or sensors. It could also mean the opacity range of negatives or the reflectance range of prints.

The minimum signal can be thought of as *read noise*, where read refers to observation or measurement. In science, *noise* is defined as fluctuations that accompany signals of interest. Noise is random and usually undesirable. Most people will think of static in a radio signal, but the "snow" in a TV picture is also noise. The signal-to-noise ratio (SNR) is a useful parameter for defining the quality of a signal or an image. It might be thought that the maximum SNR ratio would be the same as the dynamic range. For digital cameras this not the case, though both quantities depend on the FWC. I will return to this point later.

The weakest image that can be discerned on photographic film is one that just exceeds the fluctuations resulting from the grain. *Grain* is nothing more than the silver particles or clouds of dye molecules that make up the image, and *graininess* is our subjective assessment of noise in the image. On the other hand, granularity is a numerical value that is determined with a microdensitometer. Root-mean-square (rms) fluctuations are computed for the fluctuations in optical density measured with a 48 μm aperture, and the result is usually multiplied by 1000 and reported as the granularity. Figure 16.7 illustrates grain and pixel noise in images (a) and (b), respectively. The horizontal scale for the densitometer trace is greatly expanded for image (a). The dynamic range of the film could be defined as the greatest OD that can be measured divided by the rms fluctuations.

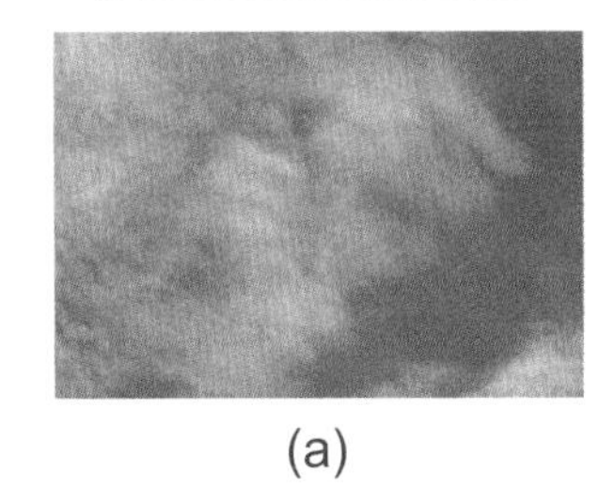

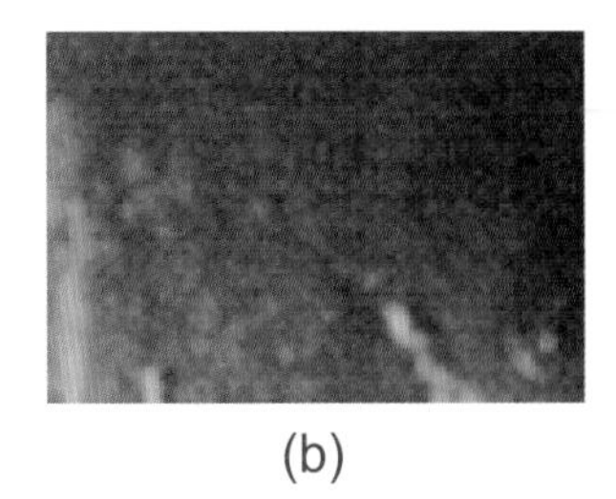

FIGURE 16.7. Noise in images: (a) film grain (simulated) an (b) pixel noise. The OD plot is typical film grain as shown in (a).

Now let's look at sensors. The maximum signal is determined by the FWC. Therefore, the dynamic range is defined as the FWC divided by the read noise. The read noise contains all of the on-chip noise that is independent of the photon signal. This noise results from the readout amplifier as well as thermal electron generation (dark noise). Dark charge accumulation can be reduced by perhaps a factor of 10 by cooling a CCD sensor by 10°C. The average dark count can, of course, be subtracted out, but the dark noise cannot. CMOS sensors have more on-chip components and more sources of noise in addition to dark noise. There are reset transistor noise (error), amplifier noise, photodiode leakage, etc. However, the dynamic range can be computed the same way.

As the photon signal increases, it soon overwhelms the chip noise, but it brings with it significant statistical noise. There is an average arrival rate for photons, but the arrival time of a photon is independent of the time since the last arrival. This is similar to the arrival of rain drops on a leaf or patrons at a box office. The distribution of arrival times is described by the Poisson distribution, and the important conclusion for this application is the following: If the mean number of events (arrivals) is equal to n, then the number of arrivals fluctuates about the mean value with a standard deviation equal to $\sigma = \sqrt{n}$. If an exposure is made of a uniform-gray card with a digital camera, the pixel potential wells will contain some average number of electrons, say n. Therefore, the signal is proportional to n while the noise is proportional to $\sqrt{n}$, and the SNR is just $\left(n/\sqrt{n}\right) = \sqrt{n}$. As a concrete example, suppose that the FWC is 20,000 electrons and the read noise is 10 electrons. The shot noise associated with the FWC is $\sqrt{20{,}000}$ electrons and the SNR is 141, while the dynamic range is 20,000/10 = 2,000. In fact, the true SNR would be slightly less since I have only taken into account the photon shot noise. Also, the dynamic range, while useful for comparison, is not a practical number, since a signal having the magnitude of the read noise could not be distinguished. A recognizable signal would require a SNR of about 3.

It is quipped that film is digital and sensors are analog. There is some truth in this because film is either "on" (black silver grain) or "off" (transparent), while the pixels in sensors store various amounts of charge that can be converted to a voltage and later digitized. This brings us to the question of how many bits are required for the digitization. As an example, consider the Canon S70 compact camera. Roger N. Clark analyzed the camera and reported that for ISO 50, FWC = 8200 electrons and read noise = 4.1 electrons [Clark 09b]. Therefore, the dynamic range is 8,200/4.1 = 2000 and the maximum SNR is $\sqrt{8200} = 91$. The digitization must adequately cover the dynamic range of the sensor, and that should determine the number of bits required. The number of steps provided by digitization increases in the procession 2^n, where n is the number of bits. Starting with $n = 8$ and sticking to even numbers, we find the sequence: $2^8 = 256$, $2^{10} = 1024$, $2^{12} = 4096$, $2^{14} = 16{,}384$. It appears that 12-bit digitization is completely adequate. The extra bit-depth is actually very useful when images are processed and mathematically manipulated to selectively enhance brightness and contrast. With marginal digitization, image processing can result in gaps in the histogram of frequency versus luminosity.

The Canon S70 shows high image quality at ISO 50, but at ISO 200 the number of electrons for the maximum signal drops to 2,150. This means that the 12-bit full scale requires amplifier gain of 2.0 for the signal and the noise with loss of image quality. Also, the dynamic range drops to 670 and the maximum SNR is only 46. Anything above unit gain, which happens at about ISO 100 for the S70, is counterproductive. Unfortunately, the higher ISO settings on point-and-shoot cameras tend to produce unacceptably noisy images. In marked contrast, the Canon 1D Mark II reaches an amplifier gain of 1.0 at ISO 1300. The Mark II (pixel pitch 8.2 μm) is thirteen times as sensitive as the S70 (pixel pitch 2.3 μm), and this difference appears to result almost entirely from the difference in pixel areas. In fact, the ratio of areas ($8.2^2/2.3^2$) is slightly less than 13.

I end this section with some estimates of the dynamic ranges associated with various imaging systems. These numbers have large errors because estimates of noise and visual quality are subjective, and technology continues to improve. As before, one stop represents a factor of 2 and n stops means a factor of 2^n:

- Sensors (DSLR): 10–14 stops (RAW readout), 8–9 stops (JPG readout)
- Color print film: 7 stops
- Color slide film: 5 stops
- Human eye: 5–7 stops (fixed retina, without iris change)

Of course, much higher dynamic ranges can be captured with digital techniques by combining several images taken with different exposures. This is the basis of the high dynamic-range (HDR) technique that was demonstrated in Chapter 13. The human eye also uses a type of HDR method when it supplements the capabilities of the retina by varying the diameter of the iris in response to the brightness of a scene.

16.7 Signal Processing and Data Manipulation

Images right out of the camera—JPG and RAW. Essentially all digital cameras permit images to be displayed and transferred in the JPEG (JPG) format. This format was developed by the Joint Photographic Experts Group and approved in 1994. It is extremely convenient for users because image processing is done in the camera, and the images are delivered ready for viewing or printing. Furthermore, JPEG is a compressed file format that reduces the memory requirements for storing images. The images are compressed by a user-selected factor and then stored. The stored images can then be reconstructed for viewing. File-size reduction by a factor of 10 yields an image that is hard to distinguish from the original, but reduction by a factor of 100 produces obvious artifacts.

Unfortunately, JPEG encoding has drawbacks, and it is not the best choice for high-quality photography. Here are the major problems:

- The color balance, contrast, sharpness, saturation, and JPEG quality must be set in the camera before the photograph is taken.
- The image is encoded with only 24 bits total, or 8 bits per color.
- Compression is "lossy" and always reduces image quality.

In the JPEG image all the settings are locked in. Of course, post-processing with a program such as Photoshop can make some corrections, but it is very difficult, if not impossible, to obtain true color balance if the original setting was incorrect. Furthermore, the 8-bit image does not respond well to multiple post-processing steps to enhance image brightness and contrast. For example, 8-bit digitization may have been good enough to describe the distributions of electrons in the pixel wells in the sensor, but a processing step can stretch the brightness range so that gaps appear for signals with only 8 bits of information.

The way around these problems is to select RAW format rather than JPEG. RAW is available on some point-and-shoot cameras and all DSLRs. In principal, RAW captures all of the information from the sensor pixels at the maximum resolution of the A/D converter, usually 12 or 14 bits. At this level, each pixel is still associated with a single color, usually R, G, or B. Therefore, a conversion program is necessary for demosaicing, sometimes called "developing," the image to obtain three-color pixels. Such programs are available from camera vendors as well as in general-purpose photo-processing packages. Camera RAW (ACR) in Adobe Photoshop is an example of a program for "developing" RAW files. The real advantage here is that the conversion programs permit the user to set the color balance, sharpening, contrast, and even, to some extent, the exposure after the fact, taking advantage of the full dynamic range encoded in 12 or 14 bits. The converted file can then be saved in a standard format such as the lossless TIF, and the original RAW file is unchanged. The RAW file acts like a "digital negative" for archival purposes.

But are RAW files really unprocessed data? First of all, we know that they are compressed, but we hope the compression is lossless or almost lossless. There is also the question of how camera manufacturers deal with bad pixels (stuck and "hot" pixels). Those pixels, that inappropriately add to the recorded image, produce fixed pattern noise which can be removed by means of dark-frame subtraction. In this procedure photographs are taken with the lens cap on and then with the lens cap off. The dark frame image is subtracted from the normal image to remove the contribution from the unchanged noise pattern. It is also possible that the camera has been programmed to do some automatic image manipulation to "fix" pixel errors. This is usually not of concern, but astrophotographers report that some Nikon cameras appear to remove the images of stars as part of automatic pixel clean-up even when preparing RAW files (see [Covington 07]).

16.8 Artifacts Resulting From Inadequate Periodic Sampling

As a concrete example of periodic sampling, consider the Canon 40D camera. The sensor contains 10.1 megapixels arranged as a rectangular array with a width of 3888 horizontal pixels in 22.2 mm and a height of 2592 vertical pixels in 14.8 mm. It is common to refer to this sensor as having a resolution of 10.1 megapixels. This is incorrect! *Resolution* is the ability to distinguish nearby points or lines, and it depends on one-dimensional analysis. For the Canon 40D the vertical resolution should be related to the pixel density in the vertical direction, and in general the resolution is proportional to the square root of the sensor size when sensors with the same aspect ratio are compared.

The actual resolving power of the sensor is somewhat less than might be thought simply by reading the pixel density. First, there is the problem that the sensor samples the image in a periodic manner, and this can lead to artifacts if the sampling rate (samples/mm) is less than twice the highest frequency (lines/mm) in the image. This is an example of the Nyquist theorem, which will be considered in more detail in the next chapter. Here it suffices to say that when the frequency of sampling is too low, there is an apparent down-shifting of frequencies in the image, an effect known as *aliasing*. A consequence is the appearance of moiré fringes or false periodic patterns. It might be thought that this problem would only appear when periodic patterns of lines, such as weave in fabrics or bricks in walls, are photographed. However, all images can be described as sums of periodic patterns with different frequencies (lines/mm), amplitudes, and orientations. For example, sharp edges can only be described accurately if high frequency components are present.

It can easily be verified that moiré fringes usually do not appear in images taken with digital cameras. More often, the fringes appear on a computer LCD when small images are displayed, because that does lead to under-sampling without correction. So how do the camera manufacturers avoid the fringes? The universal cure is to filter the image that reaches the sensor to remove the high frequencies. That's right, the images are artificially blurred. This can be accomplished in just the right amount with thin layers of a birefringent optical material (usually lithium niobate) arranged to split an incident beam into four closely spaced beams. It seems a shame to reduce the resolution provided by a high-quality lens to avoid sampling errors, but there is no alternative. When manufacturers fail to provide adequate filtering, fringing effects can be detected in resolution tests. The sensor tests reported on dpreview.com sometimes note the appearance of moiré fringes. For example, fringing problems were flagged when horizontal resolution was determined for the Sony DSLR-A700 and the Olympus E-3 cameras.

16.9 Gamma Correction

Color management is an important part of digital photography because we all want our images to look the same regardless of whether they are displayed on a computer, projected onto a screen, or printed. In a RAW image the pixel values are proportional to the exposure, and after RAW conversion the numbers are still approximately proportional to the exposure. The human eye has a logarithmic response to light, and computer screens usually follow a power-law dependence. In other words, on computer displays the relative intensity I is related to the relative input voltage V as follows:

$$I = (V)^{\gamma}, \tag{16.1}$$

where both I and V are measured on a scale between 0 and 1 and the exponent, γ, is usually about 2.2 (see [Koren 09]). This response is shown by the lower curve in Figure 16.8. Therefore, a mid-range voltage of 0.5 is represented by the relative intensity $(0.5)^{2.2} = 0.218$. This transformation leads to dark images, and a correction is clearly called for. One way to take care of this response is to apply a "gamma stretch" to the input voltage to obtain a new input voltage equal to $V^{1/\gamma}$. The voltage with gamma stretch is shown by the dotted curve. The exact correction depends on the color space selected such as sRGB or Adobe RGB. Also, some computers and video cards have built-in correction.

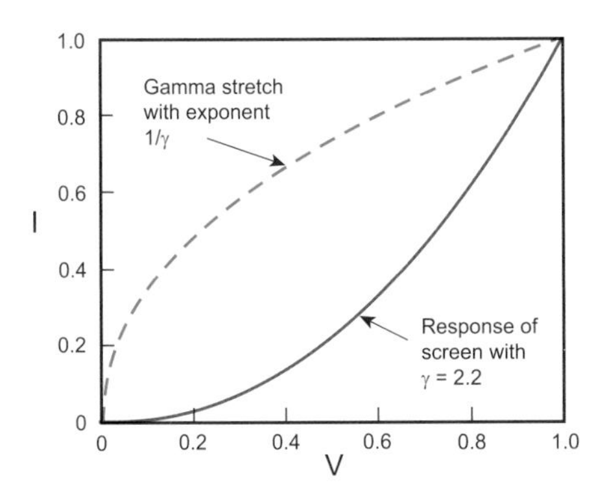

FIGURE 16.8. Intensity (I) versus the input voltage (V) for a typical computer screen with $\gamma = 2.2$. The "stretched" input voltage is shown in the dotted curve.

Serious photographers often use commercial calibration systems for their computer monitors and printers (see [Koren 09]). For example, there are units such as ColorMunki and Spyder3 that use spectrophotometers or colorimeters to monitor the brightness of the CRT or LCD screens for various RGB combinations.

16.10 Does the Sensor Really Determine the Resolution of My Digital Camera?

If everything else is perfect, yes, the sensor determines the resolution. Actually, resolution is such a complex issue that a separate treatment is required of the things that determine resolution and how resolution is measured. Chapter 17 is devoted to this subject. Also, relevant information about resolution and pixel density can be found in the camera reviews on dpreview.com. For the series of Canon DSLRs starting with the 10D, the measured vertical resolution is proportional to the number of pixels (in the vertical direction). In fact the resolved number of lines/mm is about 80% of the pixels/mm. This is remarkable, since the observed resolution is expected to be reduced by the antialiasing filter and the averaging effects of the demosaicing step in image-processing. Resolution is an elusive thing, and, as I show in the next chapter, it cannot be adequately expressed by a single number.

Further Reading

M. R. Peres (Editor). *Focal Encyclopedia of Photography, Fourth Edition.* Amsterdam: Elsevier, 2007. (See the section on digital photography.)

CHAPTER 17

What is Perceived Image Quality?

The practical effect of high-resolution technique is less that of increasing quality as reducing degradation.

—JOHN B. WILLIAMS

Psychophysics: The behavioral study of quantitative relations between people's perceptual experiences and corresponding physical properties.

—STEPHEN E. PALMER

17.1 Introduction

Everyone wants to be able to make good, sharp photographs, and most of us immediately recognize whether an image is sharp. Also, I have noticed that judges in photo competitions often say that they are checking for "critical sharpness." But what does that really mean? We have to look deeper and define some terms. What photographers are really concerned with is *image quality* (IQ) , a feature also known as *image clarity*. That is to say, how closely the details in the image match those in the original scene? The assessment of IQ, of course, depends both on the characteristics of the image and the perception of an observer. This leads to additional descriptors that are the subject of this chapter. The list, I propose, goes as follows:

- Image quality = clarity
- Resolution = spatial resolving power
- Visibility = contrast
- Sharpness = acutance (edge contrast)

Our goal is to maximize perceived IQ, and our task is to find how the other items on the list affect this perception. This is important to photographers, because we need to know how the features of lenses and sensors affect our images. We also must make decisions about resolution settings for printing images of various sizes, and beyond that there are questions about how much resolution is needed for computer monitors, digital projectors, and HDTV. What we show in this chapter is that resolution cannot be defined satisfactorily by a single number, and, furthermore, that maximum resolution is not the most important factor in determining IQ.

I will first consider the factors that contribute to the degradation of images and how those factors compound. I will then describe in some detail how the performance of cameras can be measured and presented. This part is purely technical. The image is projected by a lens and captured by a sensor. The image may also be processed after capture to enhance its appearance. I will then consider factors that depend on properties of the eye and the wiring of nerve cells in the retina and the visual cortex. I will show how quantitative information about image resolution and contrast can be combined with factors that influence human perception to help us make decisions about photographic equipment and its use. Each of us is different, of course, and we do not see things exactly the same way. Therefore, the choices will be personal.

17.2 Image Blurring and Softness

Suppose that we photograph a distant point of light, perhaps a star or an illuminated pinhole. The ideal image would have no width at all and might expose only a few pixels on the sensor. In the real world, lens aberrations, diffraction, and a host of other effects contribute to a much larger area of exposure. This distribution of intensity is called the *point-spread function* (PSF). Some point images and their PSFs are shown in Figure 17.1. Before we attempt to characterize the PSF, it is useful to list some major causes of image blurring. Here they are in what I believe to be a likely order of importance:

1. Improper focus
2. Camera or subject motion
3. Low signal-to-noise ratio
4. Diffraction broadening
5. Lens aberrations
6. Insufficient resolution of sensor or film

For some readers the order may be very different because of the care they take in making images. It is important to note, however, that the first three or maybe four items on the list depend on actions taken by the user; in other words, they are subject to user error.

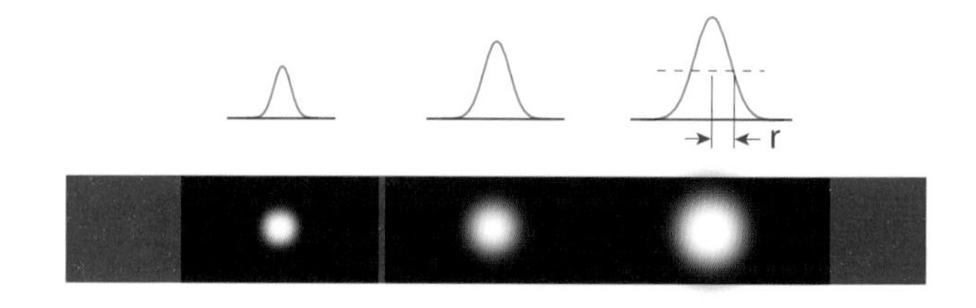

FIGURE 17.1. Blurred images and the associated point-spread functions.

We know that the PSF will always depend on more than one factor, and it is important to understand at least qualitatively how the various contributions affect the observed point spread. Taken separately, each item in the list broadens the image in a particular way, and the only one that is easily characterized is diffraction broadening. However, for present purposes it is sufficient to approximate each broadening function by the well-known bell-shaped error curve, also known as the Gaussian function.[1] Each Gaussian is characterized by a radius, r (see Figure 17.1), that I define as the half-width at half-height (HWHH). For example, suppose that an image is broadened by both diffraction (radius r_1) and lens aberrations (radius r_2) to produce an observed PSF with radius R. The relationship between r_1 and r_2 and R in general depends on the particular broadening function (lineshape function) chosen, but in the case of the Gaussian function, it is given by

$$r_1^2 + r_2^2 = R^2. \qquad (17.1)$$

If the PSFs for both diffraction and lens aberrations have radii of 10 μm, when both are present the combined radius is $\sqrt{10^2+10^2}$ µm, or about 14.14 µm. It is more instructive to consider the case where one contribution is much greater than the other. If the radius for diffraction is 10 µm and only 4 µm for aberrations, the combined radius is 10.8 µm, which is clearly dominated by the larger contribution. The lesson is that image degradation tends to be dominated by one major contributor, and that IQ cannot be improved very much by minimizing the minor contributors.

Now we are in a position to consider the broadening mechanisms one at a time. Let's take the list from the top.

17.3 Focus

Modern lenses usually autofocus, but that does not necessarily mean that the desired objects will be in focus. The user may have to set a focus point or first focus and then recompose, depending on the camera. Just as important is the depth of field (DoF) and the way it depends on camera position and settings. As a rough rule of thumb, DoF depends directly on the F-number (N) and inversely on the magnification (m) squared, i.e., DoF is proportional to N/m^2 (see Chapter 12). This means it is very difficult to get nearby objects, where m is larger, in focus at the same time. Photographs of the faces of cats often turn out better than the faces of long-nosed dogs because of the difficulty of getting both the dog's eyes and nose tip in good focus.

17.4 Camera Motion

Here again, modern technology comes to the rescue with image-stabilization built into lenses or camera bodies. This helps, but does not completely solve the problem. It simply means that we can get away with somewhat slower shutter speeds, perhaps two or three stops, when handholding a camera, but it does nothing at all for subjects in motion. In the absence of image-stabilization, the general rule is that the shutter speed in seconds for 35 mm film and FF sensors should be at least as short as 1 divided by the focal length of the lens measured in millimeters; in other words, $1/f$(mm) s. For cameras with crop factors, the denominator must be multiplied by the crop factor. For example, a Canon 40D camera (crop factor = 1.6) with a 100 mm lens should have a shutter speed at least as fast as 1/160 s for handheld photographs.

Serious photographers who want to be able to print big enlargements use a sturdy tripod along with a cable (remote) shutter-release or self-timer. When using single-lens-reflex (SLR) cameras with moving mirrors and mechanical shutters, they use the mirror-lockup feature and when possible take precautions to minimize the effects of shutter vibrations. Those who attempt astrophotography know how difficult it is to get high-resolution images of the moon and planets even when using sturdy tripods and good techniques. It should be noted that the PSF is not adequate for describing the effects of camera or subject motion, which may smear an image in one direction so that points of light become lines. There are image-processing tools that can correct, to some extent, for unidirectional motion blur.

17.5 Signal-to-Noise Ratios

It is possible that intensity fluctuations in luminance or color from pixel to pixel will be severe enough to affect image quality. With small point-and-shoot cameras this is, in fact, likely when the higher ranges of ISO sensitivity are selected. With any digital camera, noise can become a problem with underexposed images, especially when there is an attempt to brighten the dark areas with image-processing tools. The problem is that the number of electrons captured by some pixels in the sensor is so small that fluctuations in numbers are unavoidable. To minimize this problem, photographers should avoid high ISO numbers and underexposed images. For those who use histograms on their digital cameras to monitor exposure, this means "exposing to the right" so that the intensity distribution is shifted as far to the right as possible without clipping at the right end.

17.6 Diffraction

The diffraction phenomenon was discussed in Chapters 5 and 11. Basically, when a point of light is imaged using a lens with an F-number equal to N, the result is a pattern of light on the sensor that

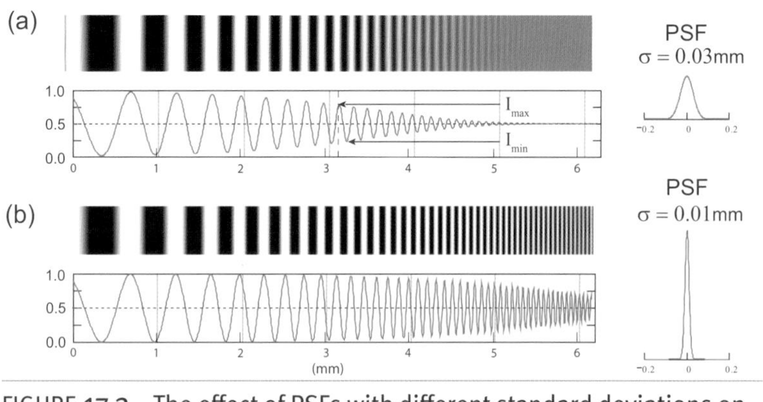

FIGURE 17.2. The effect of PSFs with different standard deviations on the contrast of test patterns for (a) $\sigma = 0.03$ mm and (b) $\sigma = 0.01$ mm.

consists of a bright center spot (Airy disk) encircled by a dark ring, a faint bright ring, and so on, becoming ever weaker. The diameter of the first dark ring is $1.35N$ μm for green light, and this is a big deal since the pixel width in some small cameras is only about 1.5 μm. So there is an incentive to keep N small, which is why compact cameras limit N to $f/8$ or smaller. In some cameras, at maximum telephoto zoom where small values of N are not available, the resolution is always limited by diffraction. Of course, there is a trade-off between diffraction and DoF. This problem was discussed in Chapter 11, and it was suggested that significant diffraction effects could be avoided by making sure that N is less than the diagonal dimension of the sensor in millimeters divided by 2. For example, a small camera with a 1/2.5 in. sensor where the diagonal dimension is 7.18 mm should always be used with $N < 7.18/2$, or 3.6. This computation assumes that prints will be viewed from distances equal to or greater than their diagonal dimension.

17.7 Lens Performance

Finally, we come to factors that are beyond the control of the user. Of course, a lens may perform better at some F-number settings and some focal lengths, but still our options are restricted. So how can we evaluate the performance of a lens? Basically, we need to know how completely the image replicates the details in the original scene. Traditionally, a lens-resolution chart with black and white line pairs (lp/mm) or sinusoidal patterns (cycles/mm) would be photographed to determine the maximum number of lines pairs/mm or cycles/mm that could be distinguished on film or a sensor. This idea is illustrated in Figure 17.2 with simulated images of a sinusoidal intensity pattern where the frequency in cycles/mm increases to the right. In part (a), the lens introduces a PSF with a standard deviation (σ) of 0.03 mm that corresponds to a radius of 1.177σ. This broadening effect causes the more closely spaced lines on the right to overlap so that the contrast is reduced. In part (b), the standard deviation of the PSF is smaller, and the higher frequencies are less attenuated. The important point is that the highest-observed resolution is always associated with the lowest contrast. We see immediately that a single number is not very useful in specifying resolution. We need to know what contrast or visibility is associated with each resolution. The conventional way to display this information is to plot the contrast versus the resolution (cycles/mm for sinusoidal patterns or lp/mm for square wave patterns) for a lens at a particular F-number. This defines one form of the modulation transfer function (MTF).

MTFs are a very important means of characterizing lenses, and photographers need to understand them. For the red intensity plots in Figure 17.2 one can compute the contrast at each frequency by measuring the maximum and minimum intensities and taking the ratio $(I_{max} - I_{min})/(I_{min} + I_{max})$. This is shown in part (a) by the intensity measurements near the 3 mm position where the frequency is 6 to 7 cycles/mm. The same information is contained in the shape and width of the PSF as in the contrast-versus-frequency measurements. This is easy to understand because the

PSF is a signature of the ranges of spatial frequencies that the lens system can pass. A sharp, narrow PSF indicates that high frequencies are present, and a mathematical (Fourier) transformation can quantify the contributions from all of the frequency components.[2] The analysis of lens performance makes use of the fact that the PSF, the line spread function (LSF), and the MTF are all related.[3]

Consider a good lens with very small aberrations so that the images at *f*/8 and *f*/22 are blurred primarily by diffraction. The MTF's that characterize the lens at these apertures are shown in Figure 17.3. All MTF curves begin at 100% contrast at low frequencies and eventually reach zero contrast at high frequencies. MTF curves exhibit a wide variety of shapes, so a given point such as 60% contrast at 100 cycles/mm does not define a unique MTF curve. The area under the curve turns out to be a better indicator of IQ than the maximum frequency reached (see [Crane 64]).

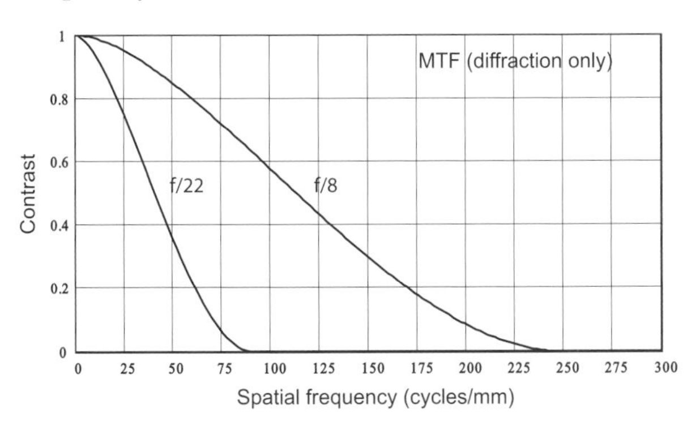

FIGURE 17.3. MTF curves for a diffraction-limited lens at *f*/8 and *f*/22.

An MTF clearly contains a lot of information about resolution and visibility, but that is just the beginning. In principle, the MTF must be measured not only for a wide range of F-numbers and focal lengths for zoom lenses but also in two orientations at representative points in the image plane. At the center of the image these patterns can have either vertical or horizontal orientation. The difference, which results from astigmatism, is usually small and the average can be taken. Significant differences are expected to appear when one moves away from the center (the optical axis). Test patterns are usually oriented in the radial direction from the center (meridional plane) or perpendicular to that direction (sagittal plane). Similarly, the line and point-spread functions can be analyzed in those directions.

The excessive amount of information in numerous MTF plots provides the motivation for a simplified type of display, also labeled MTF. The idea is to select a single frequency value, say 30 cycles/mm, and plot the measured contrast as a function of distance from the center of the image for both meridional (tangential) and sagittal orientations. Figure 17.4 illustrates test points in a FF image. Some lens manufacturers use 10 lp/mm and 30 lp/mm patterns to emphasize contrast and resolution, respectively. Also, lenses are tested at full aperture and stopped down. A typical set of MTFs is shown in Figure 17.5 for a hypothetical, high quality 50 mm lens with a maximum aperture setting of *f*/2.0. Keep in mind that the abscissa specifies the distance of the measured point on the diagonal from the center of the sensor.

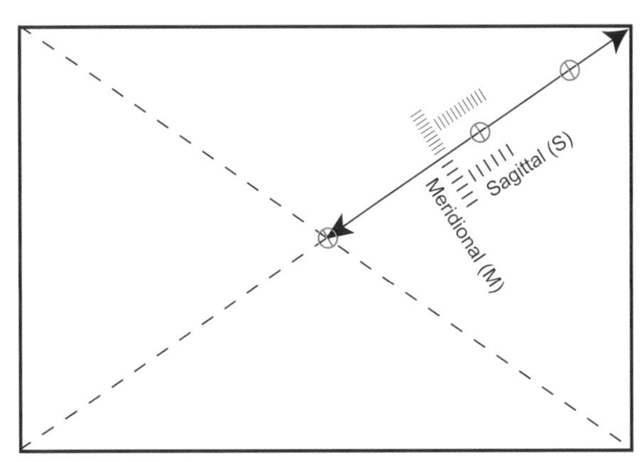

FIGURE 17.4. A sensor showing test points (red) and patterns in the sagittal and meridional directions. The length of the arrow is 22 mm for a FF DSLR.

The distance to the corner is approximately 22 mm for the full-frame sensor and only 14 mm for a sensor with a crop factor of 1.6. As rules of thumb, 10 lp/mm curves with contrast above 0.8 and 30 lp/mm curves above 0.6 represent high-quality

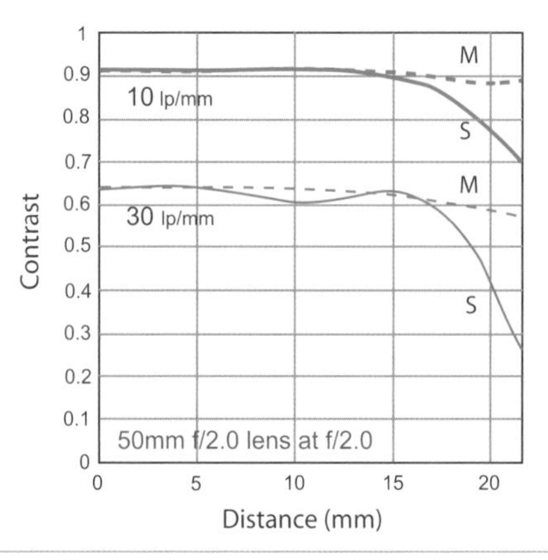

FIGURE 17.5. MTF for a high-quality 50 mm $f/2.0$ lens wide open ($f/2.0$). The Meridional line (M) is dashed and the Sagittal line (S) is solid.

images. It is much more difficult to construct wide-angle lenses that fall in this range than is the case with long-focal-length lenses. Therefore, only lenses in the same focal-length range should be compared. "Contrasty" lenses are indicated by an elevated level of the 10 lp/mm curves while the 30 lp/mm curves reveal resolving power. Also, the most pleasing representation of the out-of-focus points, i.e., the *bokeh*, occurs when the sagittal and meridional curves are close together. Most lenses are not at their best at full aperture. It is quite common to find that lenses perform better at a couple of stops smaller than the maximum aperture.

It is necessary to make two additional points. First, the line pair frequencies shown above are reasonable for the enlargement factors commonly used with 35 mm cameras and full-frame DSLRs. When smaller sensors with larger crop factors are used, the required enlargement factors will also be larger. For example, from a full-frame 24 mm × 36 mm sensor to an 8 in. × 10 in. print requires an enlargement factor of 8, while the 4/3 in. type sensor, 13.5 mm × 18 mm sensor used by Olympus would require a factor of 15 (see Table 3.1). Therefore, it makes sense for Olympus to select 20 lp/mm and 60 lp/mm for MTF plots. Also, remember that these MTF plots only take into account the performance of the lens. In order to describe the results when the lens is used with a particular camera, the contribution of the sensor must also be considered.

17.8 Sensor Resolution

The operation of digital sensors was considered in some detail in Chapter 16. A sensor samples images by means of a periodic array of detectors. The analogy between the detectors and grain in film breaks down here because grain has a random spatial distribution. The implications of this difference become clear when we recall that all images can be resolved into sets of periodic patterns with different frequencies. Consider Figure 17.6, where the black-and-white image (bar) at the top is represented by the solid black line immediately under it in the graph. This is part of a repetitive sequence of black and white bars. I performed a (Fourier) reconstruction of this shape by adding together sinusoidal functions with ever-increasing frequencies. The bottom curve contains one frequency component (1 cycle/mm), the curve labeled 2 contains two, and so on up to 20. The green curve labeled 20 is the best I can do when the highest frequency component is only 20 cycles/mm. The 20 cycle/mm component by itself is represented by the magneta curve.

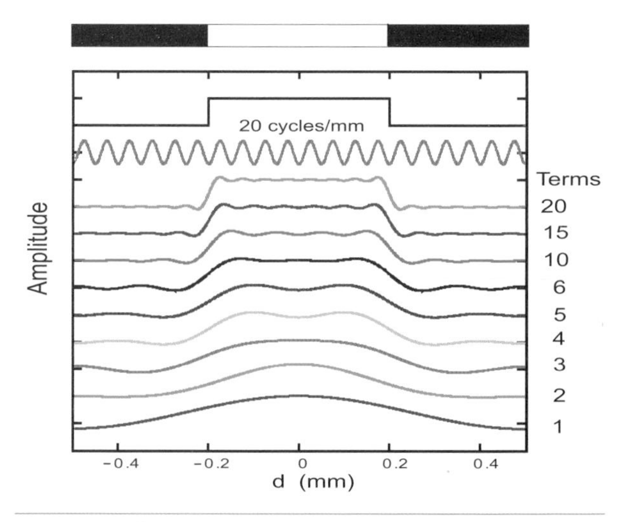

FIGURE 17.6. Fourier series analysis of a square-wave image.

It is clear that an accurate representation of this image with its sharp edges would require much higher frequencies. So we need to know what the highest frequency is that the sensor can faithfully record. The answer comes from the Nyquist theorem, which tells us that at least two samples are required per cycle (or line pair) in order to establish a frequency component. In imaging applications, the length of a cycle, the wavelength Λ, is expressed in millimeters, and the spatial frequency is specified as $(1/\Lambda)$ mm^{-1}. Therefore, according to the Nyquist theorem, the spatial frequency $(1/\Lambda)$ mm^{-1} requires a sampling rate of at least $(2/\Lambda)$ mm^{-1}. This sampling frequency is called the Nyquist limit. The components with frequencies that are too high for this sampling rate do not go away, they are just misinterpreted as lower-frequency components and added to the image. Figure 17.7 shows how this happens.

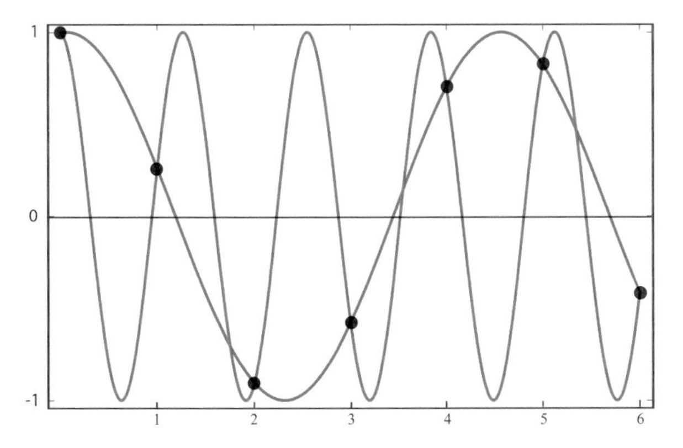

FIGURE 17.7. Illustration of the Nyquist theorem. The true frequency is shown in magenta, but inadequate sampling incorrectly identifies the lower-frequency blue curve.

The true frequency component (magenta) is sampled at too low a rate, shown by the dots, and is misinterpreted as the blue curve. This effect is called *aliasing*, and the only way to avoid it is to remove frequencies beyond the Nyquist limit from the image by means of an antialiasing (AA) filter. This intentional blurring of images was introduced in Chapter 16. The idea is to make sure that the image projected onto the sensor does not have any frequency components higher than one-half the number of pixels/mm. So this is only a problem for lenses with resolving powers that exceed the Nyquist limit of the sensors, and even with these lenses the resolution may already be reduced by aperture-dependent diffraction.

Every year the pixel count in sensors increases so that at some point sensors will out-resolve all lenses. Some camera companies have already concluded that AA filters are not worthwhile, and they leave out the AA filter altogether, e.g., Kodak DCS Pro SLR/c and the Phase One digital back. Most other manufacturers are sticking with AA filters for the current generation of cameras. The Kodak 13.7-million pixel full-frame camera is a marginal case since the sensor had only 127 pixels/mm and permits recording up to only 63 cycles/mm. On the other hand, the compact Canon G10 sports a 5.7 mm × 7.6 mm sensor with 14.7 million pixels. That sensor provides 581 pixels/mm and can record up to 290 cycles/mm, not counting the effects of the color filter array (CFA) and the need for demosaicing (Chapter 16). Not many lenses can match that.

Our concern with the Nyquist limit should not leave the impression that all frequency components below that limit are treated equally. In fact, a sensor causes some low-pass filtering because of its geometry, the efficiency of the pixel lenses, and the diffusion of charge carriers. These factors determine the MTF (contrast versus frequency) for the sensor, and the MTF for the complete photographic system is the product of the individual MTFs. An equivalent description is to say that the lens and the sensor each contribute to the overall system PSF.

So where does that leave us with regard to sensor resolution? A monochrome sensor with no CFA can perhaps be represented by a PSF with a width equal to a couple of pixels. An AA filter may or may not broaden the image by a couple of pixels, and the color filter array will insure that the pixels associated with R, B, or G light will be separated by a pixel or two. So let's take about three pixels as the minimum PSF associated with a sensor. In Chapter 11 I suggested that the circle of confusion (CoC) defining focus that is "good enough" could be defined as the diagonal dimension of the sensor divided by 1500. This seems reasonable for prints that are viewed at a distance equal to their diagonal dimension. If all prints are viewed from a standard viewing distance of about 13 in., regardless of their size, then the divisor should be scaled up by the ratio of the print width to that of an 8 in. × 10 in. print. For example, a 16 in. × 20 in. print would require a CoC equal to the sensor diagonal divided by 3000. While this is, in principal, reasonable, the practical lower limit for the CoC is about three pixel widths. With the sensor of the Canon G10, division of the diagonal by 1500 already yields only three pixel widths, and it doesn't make sense to increase the divisor. Furthermore, diffraction broadening already exceeds this CoC at F-numbers greater than 4.7, so there is not much room to maneuver.

17.9 Perceived Image Quality

So far I have considered objective measures of resolution and contrast. Now we come to the really important part. How does the image look? Attempts to quantify image quality are, of course, not new. First, there is the question of sharpness as opposed to resolution. Already in the 1960s, acutance was being advanced as an "objective correlate of sharpness." Early definitions of *acutance* were based on the steepness of edges or how quickly the intensity or the density in an image changed with distance. One measure of acutance was the square of the slope of a plot of the image density across an edge. The next step was taken by E. M. Crane of the Kodak Research Laboratories, who reported an objective method for rating picture sharpness that had been "tested in the mill of widely varied experience." His conclusion was that the perceived sharpness is proportional to the area under the MTF squared for the system when the MTF is plotted on a logarithmic axis [Crane 64]. I will return to this idea in the section on subjective image quality. At this point I turn to the popular notion of sharpening an image.

17.10 Sharpening by Means of Image Processing

Thus far, every section in this chapter has been concerned with image degradation. What about improving the image with post-processing? The image from a camera can be thought of as an ideal image in which every point has been replaced with a PSF, i.e., a fuzzy dot, having the intensity and color of the original point. All the PSFs are then added up to produce the final image. This operation is known as the convolution of the ideal image with the broadening function. A trivial illustration of this process is shown in Figure 17.8. The natural question is whether it is possible to reverse the process and *deconvolute* a real image to obtain a better approximation of the ideal image. A perfect deconvolution is not possible; but, if the shape of the broadening function (PSF) is known, the deconvolution can be approximated. Deconvolution is, in fact, the basis of some of the "smart sharpening" programs. Focus Magic is an example of a deconvolution-based program.

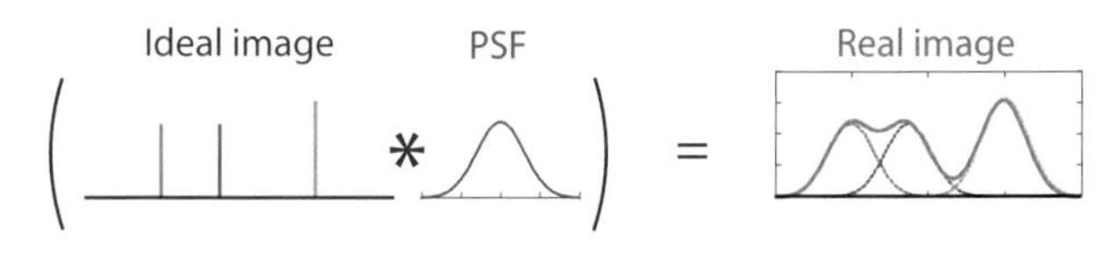

FIGURE 17.8. Convolution of an ideal image with a PSF to give a real image.

Other sharpening programs work directly to enhance edges in the image. This can be done either by means of high-pass filters or by checking the brightness of neighboring pixels. Edges are enhanced by making the pixels on the bright side even brighter and those on the darker side even darker. A blurred but sharpened edge between dark gray and light gray is shown in the image in Figure 17.9. The brightness plots show the original image (red dashed line) and two levels of sharpening with high-pass filters (black solid line and blue dashed line). It is easy to overdo this kind of sharpening and generate obvious halos around the edges.

A recommended workflow for sharpening is first to enlarge the input image from the camera to 100% (one pixel on the computer screen equals one pixel in the camera) and to apply sufficient sharpening to counteract the broadening effects of the AA filter and the characteristics of the sensor. Then, after the image has been corrected and

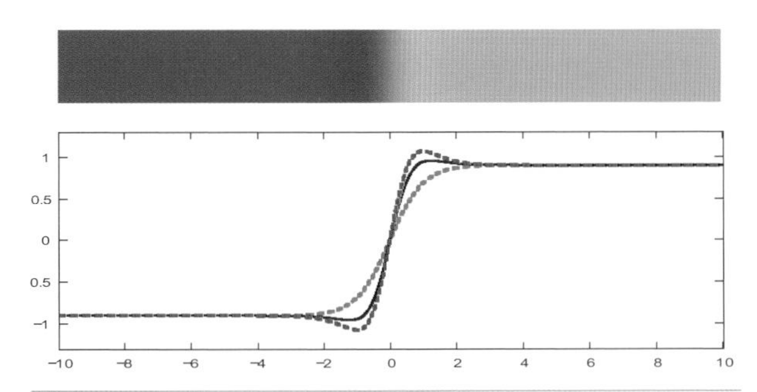

FIGURE 17.9. A fuzzy edge (red line) with two levels of sharpening by means of a high-pass filter.

the final print size has been selected, the image is resized and an appropriate output sharpening can be applied. Here it is sometime advantageous to "res" up the image to at least 350 pixels/in. before applying the final sharpening. It is convenient to use sharpening filters built into photo-processing programs, such as "smart sharpening" in Photoshop, but there are also sophisticated sharpening programs from various vendors as either plugins or stand-alones. Numerous programs including the following are available online: Sharpen Pro, PhotoKit Sharpener, and FocalBlade.

17.11 Contrast Sensitivity Function

Now we face the question of what determines our subjective assessment of image quality. Crane's finding that sharpness is determined more by the area under the MTF curve than by the highest resolution reached is a tip-off. The area measurement tends to emphasize the lower frequencies and higher contrasts. Also, physiologists have reported that human resolving ability depends strongly on contrast (see [Campbell and Robson 68]). Subjects were asked whether they could detect a grating pattern at lower and lower contrast levels. The purpose was to determine a contrast threshold below which no pattern could be discerned. The surprising conclusion was that the threshold level depends strongly on the frequency of the grating, and the lowest threshold is in the vicinity of 6 cycles per degree. The contrast sensitivity function (CSF) is defined as the inverse of the threshold, so this function has a peak in the range of 3 to 12 cycles/degree and falls off both at higher and lower frequencies. At a normal viewing distance of 34 cm, or slightly more than 13 in., this amounts to maximum sensitivity range from 0.5 cycles/mm to 2.0 cycles/mm.

Figure 17.10 provides a visual confirmation of the CSF. The grating frequency increases to the right on a logarithmic scale and contrast decreases in the vertical direction, and most people can more easily resolve full lines near the center than at the sides. Of course, the angular spread of this pattern and the viewing effect depend on the viewing distance. Also, more recent research shows that the sensitivity peak shifts to slightly higher frequencies at higher illumination levels. The MTF for the eye falls off at high frequencies because of aberrations and diffraction, and the effective MTF for the entire human visual system is expected to be the product of MTF for the eye and the CSF.

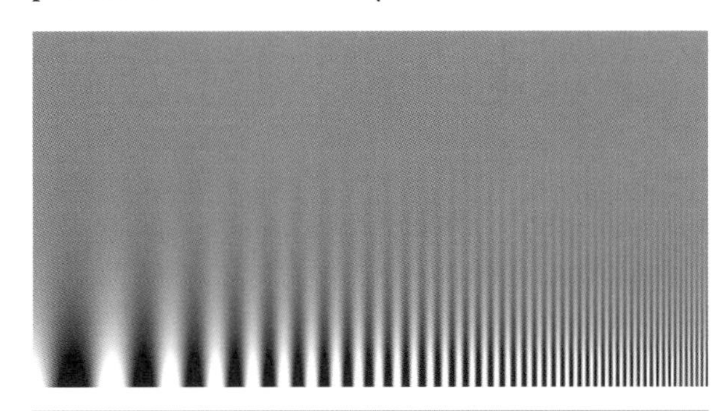

FIGURE 17.10. A plot of contrast (increasing toward the bottom) versus frequency on a logarithmic axis.

These results imply that the cells in the striate cortex are tuned to different spatial frequencies. This turns out to be the case. Ingenious experiments with gratings containing various shapes, such as line pairs instead of sinusoidal patterns, show that the (Fourier) frequency components of patterns can be distinguished by the human visual system (see [Campbell and Robson 68]). For example, the square wave in Figure 17.6 contains a fundamental frequency (one term) and higher harmonics. At lower contrasts only the fundamental frequency exceeds the threshold, and the pattern is indistinguishable from a sinusoidal pattern. At contrasts above the threshold for the next harmonic, a difference can be distinguished between line pairs and sine curves when the two patterns are repetitively switched on a computer screen. Of course, the striate cells have limited spatial extent, and they can collect only small sections of sine curves containing small numbers of cycles. Apparently, piece-wise analysis is possible with small patches of sine curves that fade out with distance. Such oscillatory patterns with bell-shaped envelopes have been described mathematically and are known as *wavelets* (see [De Valois et al. 82] and [Lee 99]).

17.12 Subjective Quality Factor

Our initial goal was to correlate perceived image quality with measurable properties of images. In 1972, Edward Granger, a senior scientist at Eastman Kodak, proposed an optical merit function that goes a long way toward accomplishing this goal. His work combined Crane's insight about the importance of the area under the MTF curve with research on the contrast sensitivity of the human visual system. What he did was to use CSF data to determine the most relevant part of the MTF for images of a certain size. Let us return to the MTF in Figure 17.3 and plot it with a logarithmic frequency axis. The resulting curves are shown in Figure 17.11.

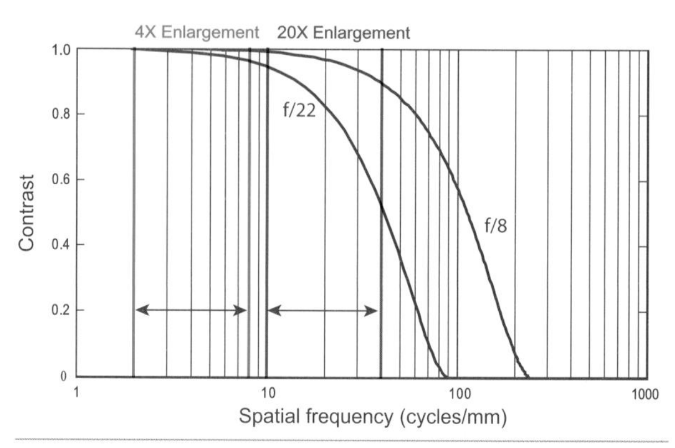

FIGURE 17.11. MTF for a diffraction-limited lens: 2–8 cycles/mm (red) and 10–40 cycles/mm (blue).

Granger defined the *subjective quality factor* (SQF) to be proportional to the area under a MTF curve (on a logarithmetic frequency axis) in the frequency range that corresponds to 0.5 to 2.0 cycles/mm on a print viewed from 34 cm (see [Granger 74; Granger and Cupery 72; Atkins 09a; Atkins 09b]). Consider, for example, 4 in. × 6 in. and 20 in. × 30 in. prints. A full-frame sensor (1 in. × 1.5 in.) image would have to be enlarged by factors of 4 and 20, respectively, to achieve these print sizes, and the required frequency ranges on the sensor would be 2–8 cycles/mm and 10–40 cycles/mm. For the 4 in. × 6 in. print, the SQF would be determined by the area between the red lines under the appropriate MTF curve in Figure 17.11. Similarly, the SQF for the 20 in. × 30 in. print would depend on the area between the blue lines under the selected MTF curve. The conclusion is that both the *f*/8 and *f*/22 curves provide similar SQFs for 4 in. × 6 in. prints viewed at 13". However, the subjective quality for a 20 in. × 30 in. print at 13 in. is lower at *f*/22 than at *f*/8 and also much lower than that for a 4 in. × 6 in. print at either *f*/8 or *f*/22. This type of quality assessment works in the real world because high-frequency components with low contrast are actually not seen.

The beauty of the SQF is that it provides a single number for evaluating a lens or optical system. Of course, the work is not quite over, because we know that the MTF curves depend on where in the image frame they are measured, and also on the orientation of the test pattern. The SQF system described here is used as the basis of the lens evaluations reported by *Popular Photography & Imaging.* (See [Granger 74; Granger and Cupery 72; Atkins 09a; Atkins 09b; Imatest 09]. [Kolb et al. 09] contains a general discussion of visual acuity including the contrast-sensitivity function.) They are able to provide a single number to describe the SQF for the complete image frame at each print size and F-number by averaging measurements at image locations in the following way: center (50% weight), 50% of the way to a corner (30% weight), and 80% of the way to a corner (20% weight). The test points are shown in Figure 17.4. The SQF numbers are adjusted to a 100-point scale and then are assigned letter grades (and colors) from A through F in 10-point steps. This is a reasonable procedure, but the user is left with no information about how quality is distributed across the image. For example, a lens with a rating of 86 for a given print size and F-number may have achieved 86 at all test sites or the scores may have been 100, 80 and 60, to give $100 \times 0.5 + 80 \times 0.3 + 60 \times 0.2 = 86$. The user may prefer center sharpness and a fall-off in quality toward the corners rather than overall softness. (See the further reading section for more on SQF, including the excellent review article by Larry White on methods used by *Popular Photography*.)

The only way to evaluate a lens, without hands-on experience, is to use several sources of information. In addition to the SQF reports and MTF curves reported by vendors, several websites now publish detailed tests of popular lenses. Perhaps the largest collection can be found at www.photozone.de. Also, http://www.dpreview.com is expanding its coverage of lenses, and Michael Reichmann often discusses his experience with interesting lenses on http://www.luminous-landscape.com. Even this level of diligence does not insure success, however, because there is still some sample-to-sample variation in lens performance. New lenses should be tested and, if found to be deficient, returned to the vendor, although there are some cases, i.e., backfocus error, where recalibration may fix the problem.

17.13 A Caveat

I close this chapter by reminding readers that everything in photography involves compromises. Now that images are digital and everyone can blow them up to 100% and beyond on computer screens, it is easy to find defects in any camera or lens. We all want large zoom ranges and large apertures with high image quality in any size print we choose to make. With present technology and materials this will not happen, although we can get a bit closer at unacceptable price levels. The new generation of super-zoom consumer lenses for DSLRs (1.5–1.6 crop factor) illustrates the point. There is a big market for all purpose, super-zoom lenses, and the major vendors have all attempted to capture it. I am talking about the 18–200 mm zoom lenses from Canon, Nikon, and Sigma as well as the 18–270 mm lens from Tamron. Each lens involves slightly different compromises, but the limitations of technology and the laws of nature are evident in the final products.

First, the maximum apertures are all modest and they shrink as the focal length increases. We find $f/3.5$–$f/5.6$ for Canon and Nikon and $f/3.5$–$f/6.3$ for Sigma and Tamron. We also observe the impossibility of uniformly high optical performance over the 10X and greater zoom ranges. With lenses having zoom ranges up to 4X, manufacturers do rather well; but that is about the limit. The present collection of lenses tends to show barrel distortion, vignetting, soft corners, and chromatic aberrations at 18 mm. As the focal length increases, the resolution and distortion change in different ways for each of the lenses. Typically resolution is best in the middle range, but the Sigma lens shows different compromises. Here the extreme focal lengths are sweet spots for resolution, while it has been reported that resolution at 35 mm is not as good. Overall, these lenses are optical marvels if one is not too critical. In fact, recommendations often hinge on non-optical factors such as focusing accuracy and efficiency of the image-stabilization system. Even if optical and mechanical parts are good, the potential buyer may be put off by the weight, the price, or the length of the warranty. Photographers must decide what compromises they are willing to make. They should also remind themselves that defects in photographs can seldom be blamed on lens quality. Good photographic techniques and an appropriate lens hood to minimize lens flare can work wonders with any lens.

Further Reading

J. B. Williams. *Image Clarity: High-Resolution Photography.* Boston: Butterworth-Heinemann, 1990. (This classic text is now dated because of its emphasis on film.)

S. E. Palmer. *Vision Science.* MIT Press, Cambridge, MA, 1999. (See Chapter 4; this book contains everything you might want to know about the performance of the human visual system, at least up to 1999.)

L. White. "Subjective Quality Factor." *Popular Photography*, November 1990, 68–75.

CHAPTER 18

The Creation and Appreciation of Art in Photography

Science and art are products of the mind;
they are of the mind yet they are the mind.

—ROBERT L. SOLSO

What people find beautiful is not arbitrary or random but has evolved over millions of years of hominid sensory, perceptual, and cognitive development.

—MICHAEL S. GAZZANIGA

18.1 What Is a Good Photograph?

To most casual camera users, photography provides documentation of people and events. Photographs "save" memories and are usually judged by their content. Did the photograph capture a likeness, was there good timing, and so on? Family and friends glance at photographs and say "that is a good one" or maybe "I'd like a copy." When the photographer has better, more expensive, equipment, there is a natural tendency to expect better photographs. When more is expected, there will be more disappointments. The photographer now wants good exposures, sharp images, and natural colors. These are technical requirements, and with some study and experience the photographer can learn what is possible with the available equipment.

At the next level we encounter amateur photographers who know about lenses, the importance of using tripods, and many technical details. They have researched cameras and have equipped themselves to make technically perfect images. Now the standards are higher, and the photographer wants approval from more than family and friends. First, there are other photographers, who are also potential competitors. Then there are camera clubs with photo competitions and experienced judges. This can be painful to the beginner because a judge might be unmoved by a favorite photograph or may dismiss a photograph out of hand because some rule of composition has been violated. Also, two judges may have very different reactions to the same photograph.

At this point, an amateur photographer may simply reject photo competitions and enjoy his or her photography as it is. This is certainly a valid option and is the one chosen by most amateurs. There are others, however, who enjoy competition and who desire feedback from other photographers. Photography is a big thing for these individuals, and they are willing to devote a lot of time and money to improving their photographic skills. They read books, attend workshops, and go on photo shoots with the aim of learning how to make better photographs. They are serious amateurs, and they may aspire to become professional photographers.

With the serious amateur in mind, I return to the role of judges. Their work is certainly subjective, but observations of many judges over time reveal some consistency. (I am ignoring truly bad judges who exhibit unreasonable biases and lack of expertise.) Some judges describe the informal guidelines they use and are willing to critique in detail images that they have judged. Here is a sample set of guidelines:

- Technical perfection (exposure, color, critical sharpness)
- Composition (pleasing layout, may involve conventional rules of composition)
- Impact (eye catching, novel, etc.)

These points may be given weights such as 30% for technical perfection, 30% for composition, and 40% for impact. What is clear is that the judge is dealing with visual art, and the process is essentially the same regardless of how the art was created. This is more true now than ever before for photographic competitions because of the widespread manipulation of digital images. Nature photography contests may impose restrictions, but some level of adjustment is usually permitted even there.

The bottom line is that the judgments are subjective, but not completely so. There are rules of composition, but they are nothing more than lists of features that either appear in or are missing from images that appeal to experienced observers. For example, an object in the exact center of an image often makes for a boring picture. One "rule" suggests that an object should be placed one third of the way from an edge of the frame, but there are many exceptions. I will return to the rules later, but here I am concerned with more fundamental questions. What is art after all, and why are certain images appealing? These are the questions that I attempt to address, but by no means settle, in this chapter.

Are the creation and appreciation of art valid topics for a science book? Absolutely! It is all about the brain. (In order to avoid confusion with terminology, let me distinguish between *brain* and *mind*. The *brain* is a physical object, and the *mind* is what goes on in the brain.) The real question is whether the neurosciences have advanced to the point that they can help us understand the creation and evaluation of visual art. This is an exciting area of research, and a lot of related work is going on. In the chapter on the perception of color, we had a glimpse of what is involved. A number of distinct areas of the brain work simultaneously on different aspects of a visual image, assigning color, detecting motion, recognizing faces, etc., and by an unknown mechanism the conscious mind experiences and understands a consistent and even beautiful image.

This chapter turns out to be an introduction to visual perception and art. I have made use of numerous recent publications on cognition and art including Margaret Livingstone's beautiful book, *Vision and Art* [Livingstone 02]. Also, Robert L. Solso's books on cognition, art, and the conscious brain [Solso 94; Solso 03] and Semir Zeki's book, *Inner Vision* [Zeki 99], deserve special mention. More recently, Michael S. Gazzaniga has addressed art and the human brain in his book, *Human* [Gazzaniga 08]. Of course, the bible of human visual perception is the wonderful book *Vision Science*, by Stephen Palmer [Palmer 99]. Finally, I note that this area of science has strong overlap with the exploding area of computational photography [Hayes 08; Efros 09].

18.2 Art and Consciousness

So, what is art? The definition of art is controversial, since there is the danger that any definition will be too restrictive. That may be, but this is my provisional definition of visual art: Visual art is a human creation that causes pleasure, and by the aesthetic judgment of the human mind it is somehow beautiful or at least stimulating. If this is too general for the photographer, just think of two-dimensional framed images that contain either patterns or representations of the world. Other definitions of art will be stated or implied in later sections of this chapter, but I do not accept the idea that all art is bound to culture. I am concerned with more universal, cross-cultural aspects of art. Also, my concept of art is closely connected with beauty, which is indeed universal. Steven Pinker points out that culture, fashion, and psychology of status (imparted by one's judgment and also ownership of art) come into play in addition to the psychology of aesthetics when judging art [Pinker 99]. I suspect that all of this is true, but I will attempt to focus on works that are aesthetically pleasing and instructive.

Art is conceived by and for the human mind. But why does our mind have this capability and inclination? It is reasonable to assume that the brain evolved for survival value, and sight evolved because of the knowledge it gives us about the world. We can speculate that visual capabilities evolved because of particular needs. For example, color assignment and color constancy aid in the selection of ripe fruits. Robert Solso emphasizes that the brain evolved to facilitate the primal functions: hunting, killing, eating, and having sex [Solso 94; Solso 03]. These activities are, of course, not limited to humans. The ability to create and appreciate art is uniquely human, however, and it may simply be a by-product of the evolution of the complex and adaptive human brain [Pinker 99]. There are strong arguments that art plays a functional role in the development and organization of the brain, making humans more effective and more competitive in a variety of ways.

By 100,000 years ago the brain of homo sapiens had reached more or less its modern capacity. Undoubtedly, people from that period and later produced decorations and carvings, but the earliest paintings that have been found date from about 35,000 BCE. These are cave paintings in the caves of southern France and in Italy. Among the oldest are those found in the Chauvet Cave, which contains about 300 murals. Less well-preserved, but possibly older, images have been found in the Fumane Cave northwest of Verona [Balter 00; Balter 08]. The Paleolithic paintings, mainly of animals, show depth and are executed in vivid colors. It was Solso's thesis that the human brain evolved to a level of complexity and adaptability that supported a high level of consciousness (awareness) and that consciousness in turn enabled art, or, conversely, that art reveals for us the appearance of conscious minds in the Stone Age.

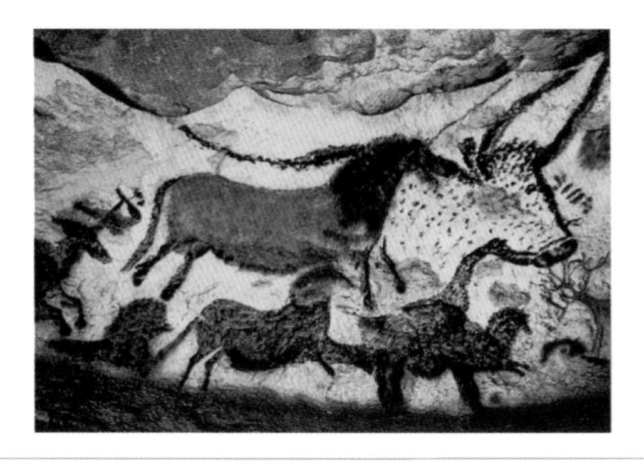

FIGURE 18.1. An example of murals from Lascaux, France (c. 15,000 BCE).

It all comes back to the capabilities of the human mind. We are blessed with a compact brain that nonetheless provides amazing power through multichannel multiprocessing. This is computer talk, which may be inappropriate. The brain is a machine that is doing some kind of information processing, but it may not be a computer. What we know is that the brain contains 100 billion neurons (nerve cells) and that each neuron is connected across synapses to as many as 100,000 other neurons. Furthermore, the connections are constantly changing in response to changing stimuli and the history of stimuli. Consequently, a mind will not respond exactly the same way to an identical stimulus the second time. Those in the artificial intelligence (AI) community, who hope to simulate the human mind, face a daunting task [Koch and Tononi 08].

Consciousness is the feature of the mind most closely associated with art, but it is also the least-understood feature. We can say a lot about the characteristics of consciousness, but we are far from understanding what it really is, or how it comes about. Consciousness is what makes us who we are. We wake up in the morning with a sense of self and experience external and internal sensations. Awareness is the essential feature of consciousness, and there is some level of consciousness even in our dream states. So consciousness is related to our global or overall mental situation, where the storm of sensory stimuli have been suitably arranged, filtered, and averaged.

We know from neurological studies of brain-damaged individuals and from functional-imaging (MRI and positron-emission) studies that the brain is organized in spatially defined modules. These modules respond to different aspects of the visual field that are mapped onto various venues of the brain. For example, the recognition of properly oriented faces is localized in the fusiform gyrus face area, and damage to this area results in *prosopagnosia*, the inability to recognize faces. The problem we are faced with is that the brain is subjected to a vast number of stimuli each second, and these stimuli are somehow processed in various locations; and yet the brain has learned how to select the essential elements and to present an understandable illusion of a three-dimensional world with colors, sounds, odors, tastes, and tactile sensations all appearing stable and natural to us. Furthermore, we can move and respond in "natural" ways in this self-centered awareness. So where does this blending and integration come from? No areas of the brain have been identified with the construction of consciousness, even though consciousness is modulated by many parts of the cerebral cortex.

Where does art fit in? One view is that art is an extension of what the conscious mind is doing for us. The mind selects essential features out of cacophony and chaos that represent a consistent view of the world, though not necessarily objective reality. We believe the mind "enjoys" understandable presentations and is disturbed by unresolved sensations. Things tailored to the capabilities of the brain attract rather than repel human beings. Similarly, effective art represents the product of millennia of experiments by artists to discover patterns and designs that stimulate the cerebral cortex. Rather than studying the structure of the brain in an attempt to discover the origin of art, artists and other experimental neurologists discover art by trial and error. The relevant structures in the brain are revealed by those compositions that are judged to be art. Feelings of joy and even euphoria are felt by the mind when "beautiful" human creations are experienced, whether sculpture, paintings, music, or theories of the universe.

18.3 How Images Are Perceived

The problem here is to understand how images received by our visual system are processed to create the world we perceive. There are two stages of perception that I will consider separately. The first has to do with automatic, fast vision capture and processing. These are processes we share with other primates. Our eyes do not make very good images. They only have reasonable resolution in the center of the visual field; and this part must be projected onto the only area of the retina that has good resolving power, namely the fovea centralis. Our vision relies on a coordinated system of extraocular muscles to orient our eyes and direct our focus to points of interest. In addition to these limitations, the amount of brain power that can be devoted to vision is not unlimited. It has long been realized that the two-dimensional images projected onto our retinas do not contain enough information for the creation of the three-dimensional world we perceive. The task of uniquely determining the features of an object from its image on the retina is an example of the inverse problem, which turns out to be indeterminate. Our visual system copes with these limitations by judicious choice of a scanning path for our eyes over the visual field coupled with perception aided by unconscious inductive inference, that is to say by statistical knowledge of prior experiences. Here *scanning path* refers to the sequence of focus points chosen to assess a scene. This overall process is known as *bottom-up-awareness* [Solso 94; Solso 03; Palmer 99].

Think of built-in defaults: just as our computers can be programmed to recognize a few letters and jump ahead to a likely word, the brain interprets images with a bias toward likely outcomes. For example, at first glance all objects are assumed to be viewed from the top and illuminated from above, and faces are quickly recognized if they have the proper orientation. Similarly, missing pieces are supplied to figures, ambiguous letters are interpreted according to context, and objects are grouped by proximity and similarity. We also get an immediate impression of depth. At close range this is based primarily on the slightly different images received by the two eyes (*binocularity*). There are also kinetic clues to distance as we move relative to objects in the visual field. Nearby objects appear to move faster than distant objects in a way that enhances our sense of depth (*motion parallax*). A major part of this effect results from the changing occlusion of distant objects. Figure 18.2 illustrates our built-in assumption about lighting with a photograph of a package of pills from the pharmacy. The right and left sides are the same photograph with different orientations, and most people interpret the right side as bumps and the left side as pits.

For distant scenes and two-dimensional representations, there are numerous (*monocular*) clues to depth. Our default is *linear perspective*, where similar-sized objects appear to decrease in size with increasing distance but still are perceived to retain their size. This can lead to errors and illusions. Also,

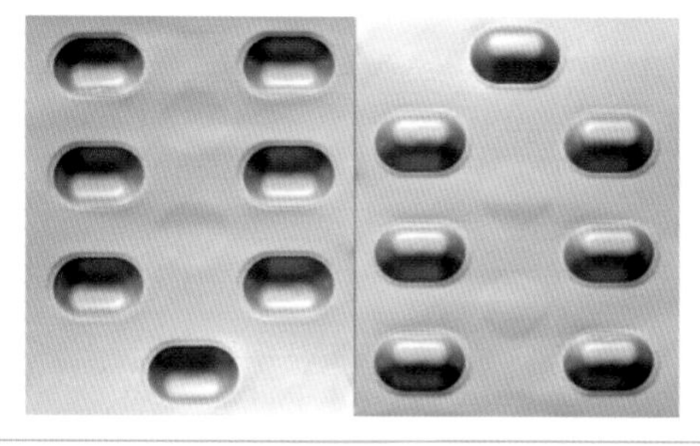

FIGURE 18.2. A photograph of a pill case with two orientations.

our natural orientation on the surface of the earth leads to the assumption that elevation is related to distance, and we note that children tend to place distant objects high in their drawings. Of course, even in two dimensions, distant objects are occluded by near objects, and atmospheric effects on clarity and contrast often distinguish distant objects. Other subconscious clues come from shadows and even the orientations of shapes and forms. A few visual illusions based on clues for distance are shown in Figure 18.3 [Zakia 02].

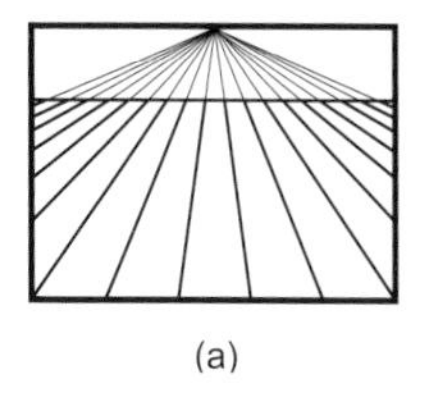

(a)

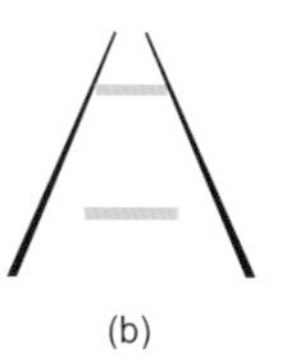

(b)

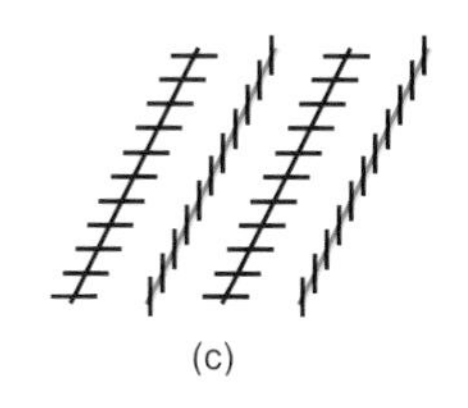

(c)

FIGURE 18.3. Visual illusions: (a) converging straight lines appear to be parallel, (b) two horizontal lines (yellow) have the same length, and (c) the red and black oblique lines are parallel.

Still other illusions are not so easy to explain. In Figure 18.4, I show the 19th century Müller–Lyer and Poggendorff illusions in parts (a) and (b), respectively, and a vase-or-faces construction in part (c). In part (a), the two lines are of equal length; the illusion of unequal lengths persists even when the arrowheads are replaced with circles or squares. In part (b), the oblique lines exactly match up regardless of how it appears. According to recent studies, both illusions can be explained on the basis of statistics of what the visual system has encountered in previously viewed scenes [Howe and Purves 05; Howe et al. 05]. Reactions to part (c) apparently depend on what the observer is prepared to see except, of course, in cases where the viewer is familiar with the type of illusion. Readers can find numerous still and animated illusions online at Wikipedia and other sites (see ["Optical Illusion" 09; Bach 09]).

Inference drives many of our perceptions and leads us to see or imagine familiar objects in the shapes we see. Contours that appear to represent one thing but then shift to a different appearance

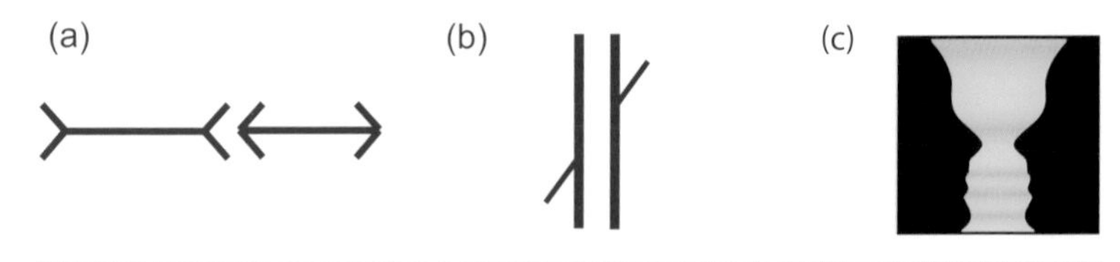

FIGURE 18.4. More visual illusions: (a) Müller-Lyer, (b) Poggendorff, and (c) Vase or Faces.

when a different side is studied provide examples. Another set of illusions results from the fact that our visual "where pathway" (Chapter 15) makes use of illuminance, or perceived brightness, rather than color in establishing position. For example, objects that have a different color than the background but the same luminance may appear to have an unstable position.

In bottom-up processing, salient features of a scene receive immediate, focused attention or fixation, and the scan path of our vision quickly samples the image and returns to the more interesting features for multiple hits. Studies of eye motion reveal that faces are high on the list of salient objects. Saccadic eye movements, which consist of angular jumps up to about 60°, each requiring 20 to 200 milliseconds, direct both eyes to a sequence of visual points. This generates a few high-resolution snapshots from foveal vision per second while the data from fuzzy peripheral vision is simultaneously received at much higher data rates. It is believed that this unbiased behavior of the visual system provides more or less the same impression of a scene to all observers.

The second stage of visual perception involves intentional direction of vision, i.e., volition. Here the "searchlight" of attention is brought to bear on features according to personal schemata. This is known as *top-down awareness*. Schemata are essentially schemes for organizing knowledge based on what is important to an individual. Training, background, and even role-playing can affect the sampling and analysis of an image. This implies a "mind set" that can lead to increased efficiency but also to automatic, "knee-jerk," assessments. At its worst, this can result in rigidity and a reluctance to accept change, the opposite of open-mindedness. For example, we see schemata in fads that dictate what styles are "in" during a given historical period. Schemata are not only locked in by experience or inclination, they may exist temporarily because of a recent work experience or even a role-playing exercise. For example, you might be asked to assume that you are a policeman while viewing a scene.

In viewing a work of art, the initial visual scan (bottom-up awareness) may emphasize faces and other salient features to provide an initial impression. The immediate impact of the work may result from bright colors or perhaps novel features. In the second stage (top-down awareness), the observer can bring to bear education, experience, and training to evaluate the work in context. At this point, discussion with others and additional observation may grossly affect the assessment of the observer, and personal schemata may be modified or strengthened. Art judges, like all of us, have sets of schemata in place. In a given period most of the judges may have similar schemata, and that poses a problem for new styles. There is also the schema of the jaded judge who is tired of images of sunrises, sunsets, hot air balloons, blurred water, and so on.

18.4 Why Do Images Attract or Repel?

Now we come to the crux of the matter. Why do we respond as we do to art? It is easy and fun to speculate. Consider the following suggested "attractants."

Representations and reminders of stimulating experiences. Representational art certainly reminds us of people and events. Food, games, combat, and attractive people are all featured in popular art. This brings to mind good documentary photography that has impact and evokes feelings. At a deeper level, all art represents nature, and some universally satisfying features of images reveal our evolutionary past. Natural landscapes are preferred, especially ancestral landscapes like the African savanna. Furthermore, images of natural objects such as mountains, clouds, and trees often reveal patterns that repeat at increasing magnification. For example, shorelines may have a similar appearance, at least statistically, when viewed from 1000 ft, 5000 ft, and so on. The complexity of these images can be characterized by a similar fractal dimension, and it turns out that observers prefer images that have the fractal dimension of nature even for things like city skylines.

Sexual stimulation, overt or subconscious. At the simplest level, images may remind the viewer of sex objects or sexual activities that are stimulating. The suggestions may be more subtle as in, for example, hidden faces or subliminal images in artworks. It has been suggested that Stone Age cave paintings contain abstract representations of vulvae. At a different level, art may serve as the plumage of the artist to impress potential mates, à la the ostrich. Similarly, the ability to appreciate or even afford art can confer status.

Curiosity and inquisitive tendencies. Animals tend to explore neighboring areas even when food is plentiful in the vicinity. This can have survival value when environmental conditions change, and curiosity appears to be an essential feature of our genetic makeup. This may explain the stimulating effect of art since art involves novelty as well as repetition. In our search for novelty we travel to new places, visit museums, and sometimes even attempt to create art. This line of thought fits well with the idea that all forms of fictional experience including art, drama, and imagination play an important roll in organizing the brain—a major adaptive task throughout life [Gazzaniga 08].

This list is intuitive and may capture important attributes of art, but it is in the category of suggestion rather than proofs. Experience shows that the brain is not only capable of creating art and judging art, but that it is inclined to commit its resources in those directions. Since the late 20th century functional imaging of the brain by neuroscientists has revealed details of the operation of the visual system and pinpointed areas of the brain that are involved in various tasks. This is very encouraging to those who are attempting to discover the biological basis of art creation and appreciation, but we are far from being able to understand anything about our emotional response to art from physical studies of the mind in action.

At this stage we must be satisfied with a descriptive approach based on cataloging the way the mind perceives and reacts to art. This is an experimental study, and through the millennia artists have unknowingly served as neurologists in determining which constructions and compositions stimulate the mind. Our museums are filled with works of art that have stood the test of time, and every day new works of art, including all varieties of photographs, are competing for approval. The take-home message is that we already have a vast collection of works of art that have been found to stimulate the mind, and this reveals to us the way the mind functions. We can use the successful features of works of art in new creations, and we are free to experiment with new art forms to determine their compatibility with the structure of the human mind. Humans have very similar visual capabilities, but our schemata are all different. This accounts for subjectivity in the evaluation of art but, in spite of that, there is wide consensus on what represents good art.

What we can learn from successful images. Just as the conscious mind thrives on the consistency and stability of our world view in spite of the plethora of sensations and the jumpy nature of eye movements, it responds favorably to simplicity in framed images. That does not mean that details should be missing, but rather that there should be a center of attention and a minimum of distractions. The eye should be directed to the most important areas by visual clues and should not be confused by multiple areas of equal importance. Similarly, framed images that are balanced and not boring are favored. Recognition of these features of successful images has led to universal rules of composition.

Beyond the simple admonitions to keep it simple and avoid distractions, we can list features that have found favor with most observers. These guidelines to composition can save time and, to some extent, substitute for experience for beginning photographers. With digital photography, where each captured image costs practically nothing, it is all too easy to shoot away without planning and later to realize that the compositions lack something important. I believe that lists of guidelines can increase awareness about composition and motivate planning by photographers. They may also forewarn about likely reactions of critics.

With the usual caveat that guidelines are not rules and that there are exceptions to all of them, I list a selection of guidelines that are frequently encountered.

- Establish a major object or area of interest. It should be relatively easy to answer the question "What is the subject of this photograph?" The object of interest can be isolated by means of placement, background, and depth of field. In some cases, natural lines and contours can be used to direct the eye to the subject. Natural frames within the image can also be helpful.
- Avoid distractions. Some major distractions are competing points of interest, such as bright areas (especially at the edges), large fuzzy areas in the foreground, busy backgrounds, and lines that run directly to a corner.
- In general, avoid placing objects directly at the center of the frame unless symmetry demands it. Center placement, especially of small objects, makes balance difficult and tends to be boring. As a corollary, avoid placing the horizon exactly in the center of a picture. The subtle difference in composition between Figure 18.5(a) and Figure 18.5(b) is sufficient for most observers to find (b) more pleasing.
- Consider using the "rule of thirds." Divide the image into thirds in both the horizontal and vertical directions by means of imaginary lines to make nine blocks. The crossing points of these lines, namely one-third of the way in from both the vertical and horizontal edges, make favorable "hot spots" for the placement of objects or points of interest in the image. Placement at a crossing point works for the geyser in Figure 18.6; however, the symmetry and size of the orchid in Figure 18.7 leaves little choice about placement.
- Give moving objects extra space in front for their anticipated movement. This is obviously desirable in most cases. In the event that the moving object has a "tail," such as the wake for a boat or the jet trail of an airplane, it may be better to assume that the tail is part of the object.
- Avoid awkward clipping of objects or features at the edge of images. For example, windows in walls should usually be totally in or out of the frame. Tight cropping to show a face or

(a)

(b)

FIGURE 18.5. Placement of objects on- and off-center.

FIGURE 18.6. Castle Geyser and the rule of thirds.

FIGURE 18.7. Orchid symmetry controls the composition.

perhaps a waist length portrait may be fine, but avoid clipping small parts. For example, try not to clip a hand or an ankle and foot.

This list simply attempts to put into words some of the features that contribute to our like or dislike of certain images. Sometimes placement is so obvious to most people that a rule is superfluous. Consider, for example, classic portraits created over many decades of faces that show two eyes. The faces are seldom straight on, and the dominant eye tends to be in the exact center of the canvas with remarkable consistency. The location of noses and mouths is much more variable. This observation is interesting but not particularly helpful.

With all this discussion of composition, one should not lose sight of the importance of subject matter. A valid criticism of any work of art is that there is nothing inherently interesting in the subject. This immediately gets us to the most subjective part of art criticism. One photographer may find art in the placement of a cigarette butt on a sidewalk or the location of a weed growing from a crack in concrete while another photographer is bored stiff by the same subjects. The choice of the subject of a photograph is quite personal. I personally am attracted to images that help me see the world in a new way. That might mean detail and colors in insects and birds, frozen action in sports, or composition and color in landscapes. The reader can insert his or her choices here as well.

18.5 How Knowledge of the Visual System Can Enhance the Artist's Bag of Tricks

According to the aphorism, "knowledge is power," and knowledge of the characteristics of the human visual system should impart some advantage to photographers and other artists. I suggest two strategies for the use of this knowledge. The first and more important one has been used by successful painters perhaps without their understanding. Thus artists have given conventional images extra impact through the use of unrecognized illusions. For example, impact can result from the use of equiluminant objects and backgrounds [Livingstone 02]. A commonly cited example is Claude Monet's *Impression, Sunrise,* shown in Figure 18.8. When properly reproduced, the solar disk and its background have the same luminance, and for many observers the sun attracts attention because our visual system has a hard time fixing its position.

FIGURE 18.8. Monet's Impression, Sunrise.

There are also illusions of motion resulting strictly from the geometry of an image, and, in particular, from closely spaced lines of high contrast. High contrast layers adjacent to equiluminant layers can impart the impression of motion. Here again Monet comes to mind for his genius at making water appear to flow in his paintings. I have put together an illustration of this effect, in Figure 18.9 where I hope the reader will experience a sense of flow in the solid areas.

Still in the category of conventional images we find images hidden within pictures that take advantage of the human propensity to recognize faces from a minimal set of clues [Solso 94; Solso 03]. A famous example is Salvador Dalí's *The Slave Market with Disappearing Bust of Voltaire* (1940). Numerous examples of this can be found online, but as far as we know they have been limited to paintings. With the capabilities of readily available photo-processing software, however, the introduction of hidden images in photographs is within easy reach.

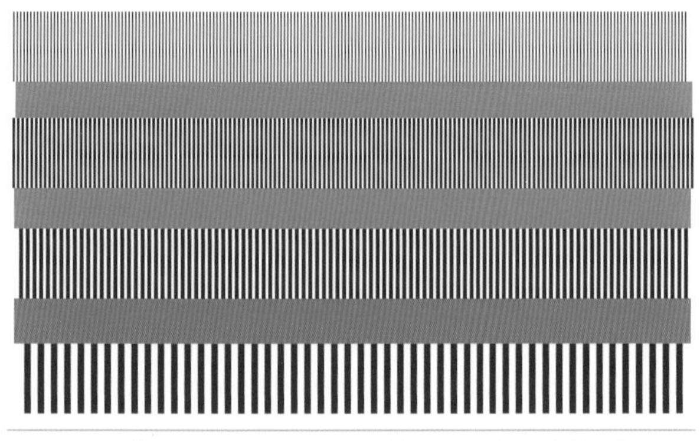

FIGURE 18.9. Impression of flow, a visual illusion.

The second category of images, suggested by the operation of the human visual system, are images that have been designed to have shock value by defying the built-in assumptions of our bottom-up awareness system (inductive inference engine) [Palmer 99]. By that I mean images that at least at first glance appear to represent impossible or extremely unlikely situations. Images that provide contradictory clues for determining depth are easy examples. In Figure 18.10, three identical images of an egret are located in a beach picture to give conflicting clues about size. The basic rule is that the perceived size of an object projected on the retina is proportional to the perceived distance to that object.

Visual dissonance is a state of psychological tension that results from seeing something that is different from what one expects to see. Functional imaging studies show that this state leads to the activation of more areas of the brain than would be required for the processing of expected scenes. This increased stimulation may lead to attraction or repulsion, but at least it is stimulating. Surreal art such as that created by René Magritte and Man Ray provides numerous examples [Meuris 04]. Figure 18.11 I show a composite photograph inspired by the work of Magritte.

Here the illusion of the moon over Sandia Peak is disturbed by a look at the dark space. Another example of dissonance results from impossible reflections in mirrors. A fanciful mirror photo of the impossible is shown in Figure 18.12.

FIGURE 18.10. Identical images of an egret in three locations.

FIGURE 18.11. Moon Space, a hole in reality.

18.6 Reflections on Art in Photography

I have tried to confront the question of why composition matters in photography and other visual art. Evidence indicates that art is closely related to and maybe inseparable from consciousness. The organization of the human mind to permit full color, binocular vision and probably general awareness as well requires years to develop. Our imagination, including all forms of art, certainly aids in the organization process. The result of this process is the mature mind that experiences pleasure and even exhilaration at suitably arranged representations of the world. We describe such arrangements as art and are especially delighted when such representations expand our understanding through thought-provoking novelty.

Our enjoyment of our colorful three-dimensional world is tempered by the knowledge that our conscious world is an illusion. This illusion replaces the flood of stimuli from our senses with a calm, stable world view. Furthermore, our conscious state attempts to minimize our visual defects and fill in unknown parts, such as blind spots, by sophisticated guesswork. This is, of course, the source of so-called visual illusions. As observers of art we are limited, both by the universal bottom-up awareness and by our own unique top-down awareness.

Art exists because the human mind enjoys stimulation by visual images and the response of our visual system is sufficiently universal that we can share the pleasures of visual aesthetic stimulation with others. Artists through the millennia and neurologists more recently have discovered empirically what types of images appeal to the human mind, and the experiments continue to our great benefit. What I have learned from this study is something about how the human mind copes with the limitations of our visual system, and, more importantly, that there is a vast world of wonderful art to be enjoyed. And last, but not least, there is still a world of wonderful art yet to be created by us photographers and other artists.

FIGURE 18.12. The author with his camera.

Historical Note on Enlargers

Photographic enlargers have existed since the earliest days of photography. The first enlarging system was described by Sir John Herschel in 1839. Since transparent negatives were not available until about 1850, all of the early enlargers were essentially large copy cameras. The problem for both copy cameras and negative-projection systems was to find bright-enough light sources so that exposures could be made in a reasonable amount of time. One solution was to use sunlight, and solar enlargers known as solar cameras were in use with some success throughout the 19th century. The exposures were typically so long that the solar cameras had to be continually adjusted to track the motion of the sun, and of course cloudy days were a problem. Also, a variety of artificial light sources were used. The most popular of which were limelight, electrical arc, and burning magnesium. Acetylene lamps were even used, but a series of explosions dampened the enthusiasm for that light source.

Even though Louis Jules Duboscq demonstrated an enlarger with an electric light (carbon arc) at the Paris Photographic Society in 1861, it was not really practical for individual photographers because of the great cost of batteries and later of steam-operated generators. Thomas Edison demonstrated his incandescent light bulb in 1879, but the time was not right for general acceptance because electricity was not generally available. It was not until about 1912 that enlargers with electric bulbs started appearing on the market for amateurs. This seems reasonable when one realizes that, by 1907, only 8% of homes in the United States had electricity. The expense and hazards of early enlarger light sources have been recounted in great detail by Eugene Ostroff in his profusely illustrated review article [Ostroff 84].

What Is Behind the Rules of Optics?

You will have to brace yourselves for this—not because it is difficult to understand, but because it is absolutely ridiculous: All we do is draw little arrows on a piece of paper—that's all!

—R. P. FEYNMAN

In science, we look for unifying concepts and general theories. Light is an electromagnetic wave, so it should be described by Maxwell's theory of electromagnetism. But now we can go beyond that description and predict light intensity and the interaction of photons with electrons by means of quantum electrodynamics (QED). QED is a model of the world that is totally unfamiliar to most of us, but it is easy to see how it works without using advanced mathematics. R. P. Feynman, one of the founders of QED, explained in detail the basic ideas and how to use them in his beautiful little book, *QED, The Strange Story of Light and Matter*. This brief appendix is an introduction to Feynman's ideas. It is thinly disguised graduate-level physics. I hope it will motivate serious science students to explore this topic through further reading.

What we want to know is how light gets from a source to a detector such as those in the sensor of a digital camera. We know it has to do with waves and their interference. Fortunately, it turns out that a lightwave can be represented by a rotating arrow (vector)—think of the hands of a clock. The important things to keep up with are the length (amplitude) and turn (angle) of the arrow. Each way that a photon can go from place to place is described by a single rotating arrow. Feynman proposed a set of rules for dealing with these arrows that permit us to compute the outcome of any optics experiment. He tells us not to worry about understanding them, just learn to use them, because they describe the way nature behaves. So here they are:

- If an event can happen in alternate ways, determine the arrows for each way and then add them by placing them head to tail. The "sum" is obtained by drawing an arrow from the tail of the first arrow to the head of the last arrow. The length of this final arrow is called the "probability amplitude."
- The probability of the event is determined by squaring the probability amplitude.

What this says is that photons take every possible path, not just those specified by Hero's law of reflection, Snell's law of refraction, etc. However, when the little vectors for all of the paths are added up and the sum is squared, there is little probability that a photon will violate the well-known laws of optics. This is known as Feynman's sum-over-path quantum theory, and it is consistent with the rules of optics and much more. The procedure, not why it works, can be understood by looking at some examples. First, consider photons of red light going from a Source at *S* to a Detector at *D* as shown in Figure B.1. The wall between the source and the detector has two very small holes (slits) that are located at *A* and *B*. The holes are so small that the normal laws of optics do not work (diffraction effect). We start the clocks when the photon leaves the source and the clock hands (little arrows) rotate at the frequency of light until the photon reaches the detector. A photon moves at the speed of light, so the orientation of the arrow will depend on the length of the path if nothing else happens along the way. Since the two paths have different lengths, the travel times and the rotation angles will be different.

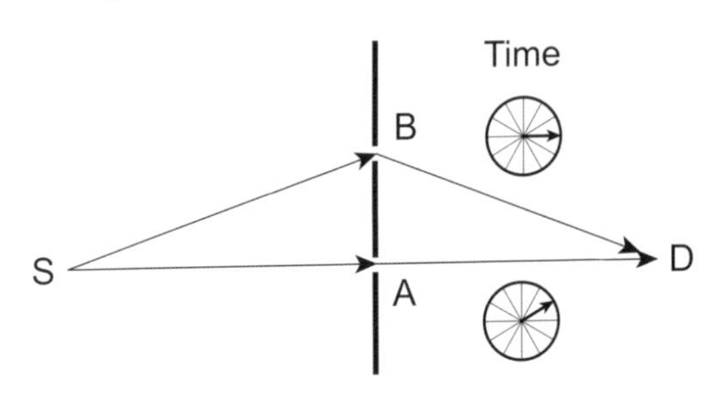

FIGURE **B.1.** The two-slit experiment.

Here is what happens. When the hole at *B* is blocked, the detector measures some light intensity, e.g., 100 photons per second. This is perhaps 1% of the photons that leave the source. The arrow representing this path has some direction, say 2 o'clock, and the length 0.1. With only a single arrow, direction does not matter, and length squared gives 0.01 as the probability of detecting a photon. If *B* is open and *A* is blocked, the same intensity is detected as when only *A* was open. The arrow for the path through *B* has a direction, say 3 o'clock, and the length is again 0.1.

Now for the good stuff. When both holes are open, the intensity can be anything between 0 and 4%, and it depends on the separation of the holes as well as the frequency of the light. The exact number must be computed from the arrows. According to the rules, we connect the arrows and determine their sum. (See Figure B.2.)

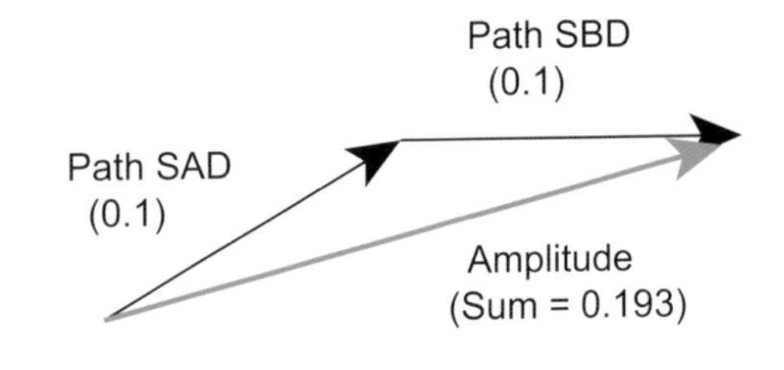

FIGURE **B.2.** The sum of arrows (vectors) for two paths.

In this case, with arrows at 2 o'clock (30°) and 3 o'clock (0°), we find that the amplitude squared is $0.193^2 = 0.037$, or about 3.7%. This procedure can be repeated for different heights (vertical position) of the detector to map out the intensity pattern for the double-slit experiment.

The reflection of light from a mirror, discussed in Chapter 6, provides another illustration of the sum-over-paths method. As shown in Figure B.3, light from the source can reach the detector by reflecting from any point on the mirror—though here we only consider a thin strip of mirror sticking out of the paper. The length of each path is given by Equation 6.2, and the time required for the journey can be obtained by dividing the path length by the speed of

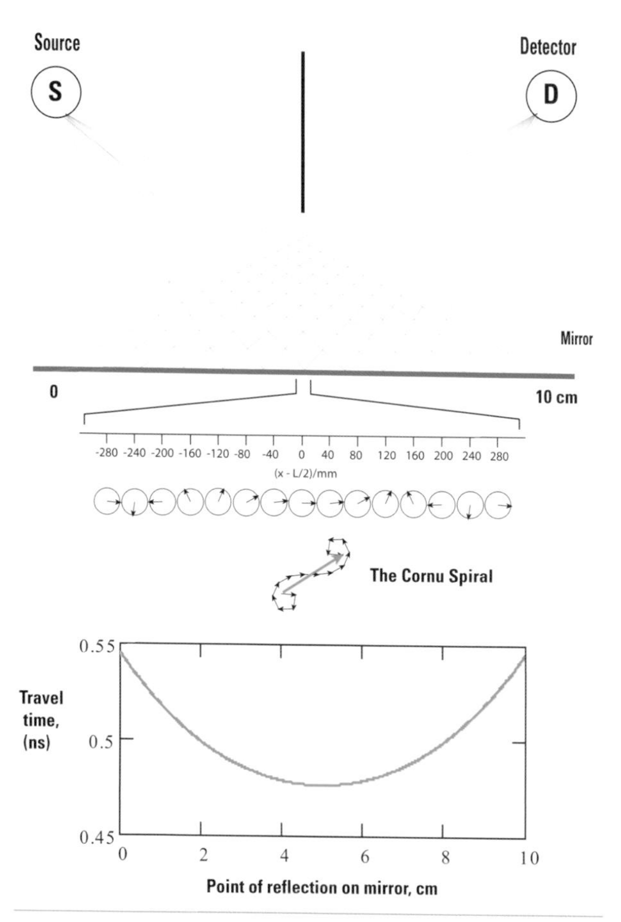

FIGURE B.3. Reflection of light with wavelength $\lambda = 550$ nm from a mirror. The length of the mirror is 10 cm and the source and detector are located 5 cm above the left and right ends of the mirror, respectively.

light, $c = 2.997 \times 10^8$ m/s. The result of this calculation is shown at the bottom of Figure B.3 for each reflection point. In this illustration, the length of the mirror is $L = 10$ cm, and a small section centered on $x = 5$ cm has been greatly magnified to show the clock angles for a region about 600 μm long. When the small arrows are added, the length of their sum is found to be determined by a few arrows in the center. It would not change significantly if more arrows were added at the ends, because they would only continue to go in circles. Note that the spiral formed by the arrows is just the Cornu spiral found in optics textbooks. So the probability is nonzero for reflection from the vicinity of the point where the angle of incidence equals the angle of exit. This agrees with Hero's law and also Fermat's principle of least time. It is, however, more general, since it correctly handles cases where the maximum-time path is chosen, e.g., reflection from a cylindrical concave mirror.

There is still more to recover from the sum-over-paths analysis of reflection. Suppose we chop off all of the mirror except the piece from $x =$ 0 cm to $x =$ 2 cm. We know that no light from that region is reflected to the detector, but it is still worthwhile to look at it in detail. If we take the angle of the arrow for reflection from $x =$ 1 cm as the reference angle and examine the other angles for reflection in that vicinity, we find the result shown in Figure B.4. Every 0.4 microns (400 nm) the orientation changes by 180°, and so the amplitude for reflection completely cancels out for that region, i.e., the arrows add up to 0. However, the diagram suggests how the reflection can be recovered. All we have to do is to scrape away the mirror at the positions where the arrows are aimed down! Actually, 3 o'clock to 9 o'clock. When we add up the remaining arrows and square the sum, we find that the probability is large for reflection of 550 nm light to the detector. What we have done is to create a diffraction grating. When white light is used as the source, each wavelength (color) diffracts strongly at a different angle as shown by reflection from the CD in Figure 2.4.

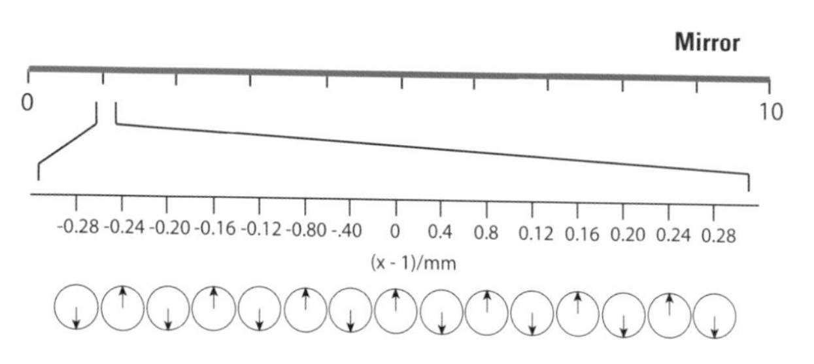

FIGURE B.4. Reflection from the end of the mirror far from the path of least time.

These examples show the origin of the rules of optics at a fundamental level including, when appropriate, Fermat's principle of least time. However, there are still some questions about the interaction of light with matter that results in reflection, transmission, and refraction. The angle for each rotating vector depends on the total transit time through a lens or collection of lenses including the time inside glass where the refractive index differs from one. Other things can happen along the way, as well, and the interaction of light with matter (more precisely photons with electrons) requires a few more rules. To handle reflection from a sheet of glass, when some of the incident light is transmitted, the arrow for the reflected light must be reduced in length so that its square equals the probability of reflection. Also, the process of reflection (from air back to air) reverses the direction of the arrow, i.e., gives a phase shift of 180°. This reversal of direction also applies to the mirror we considered above, but in that case the reversal did not affect the result since it applies to every arrow. The phase reversal is essential, however, for dealing with sheets of glass where light can reflect from more than one surface, and the arrows for paths with different numbers of reflections have to be added. Another rule, apparent from previous comments, is that light appears to require a longer time to pass through a piece of glass than through the same distance in air, and the time in glass is approximately equal to the time in air multiplied by the refractive index.

But now, the number of rules is growing, and it is not clear where they are coming from. Actually, things are not as bad as they might seem. The phase shift on reflection and the refractive index are, in fact, the results of calculations involving all the atoms in the glass. A single rule specifies the scattering of a photon by an atom. The arrow shrinks by an amount that depends on the kind of atom and also turns by 90°. I am not going to complete these calculations, but simply tell you that when the arrows for the photons, scattered from all the atoms, in the direction of reflection are added, the result is a wave at the surface of the glass that can be described by an arrow appropriately reduced in length and with a reversed direction. This same model of scattering of photons by atoms also produces a transmitted lightwave that can be described by an arrow with exactly the same orientation as would be calculated by assuming that the velocity of light inside the glass is reduced by the factor n. A detailed discussion of these calculations can be found in the further reading references.

All of this stuff about rotating arrows is just a way to approximate the final quantum mechanical wave function. Of course, quantative calculations require the summation of very many closely spaced paths and, in fact, integral calculus must be used. The simplified diagrams and computations presented here are designed to reveal the basic ideas rather than to provide tools for exact computation.

Further Reading

R. P. Feynman. *QED: The Strange Story of Light and Matter*. Princeton, NJ: Princeton University Press, 1958.

E. F. Taylor, S. Vokos, J. M. O'Meara, and N. S. Thornber. "Teaching Feynman's Sum Over States Quantum Theory." *Computers in Physics* 12 (1998), 190–199.

J. Hanc and S. Tuleja. "The Feynman Quantum Mechanics with the Help of Java Applets and Physlets." Available at http://pen.physik.uni-kl.de/w_jodl/MPTL/MPTL10/contributions/hanc/Hanc-Tuleja.pdf, 2005.

Derivation of the Lensmaker's Equation

The *lensmaker's equation* for a thin lens is easily derived from the equations for the focusing powers of the two surfaces as described in Equation (7.1). The situation is illustrated in Figure C.1, where the radius of curvature for the second surface, R_2, is a negative number.

Now imagine that the two dagrams are overlapped, so that the lines defining the central plane of the thin lens coincide and the region having refractive index n_2 is confined between the two curves. The result is that the image plane of the first surface lies on the right side of the second surface at the position q_1. Therefore, the subject for the second optical surface can be defined as $p_2 = -q_1$. The equations for the powers of the surfaces are then given by

$$\frac{n_2 - n_1}{R_1} = \frac{n_1}{p_1} + \frac{n_2}{q_1}; \quad \frac{n_3 - n_2}{R_2} = \frac{n_2}{-q_1} + \frac{n_3}{q_2}. \tag{C.1}$$

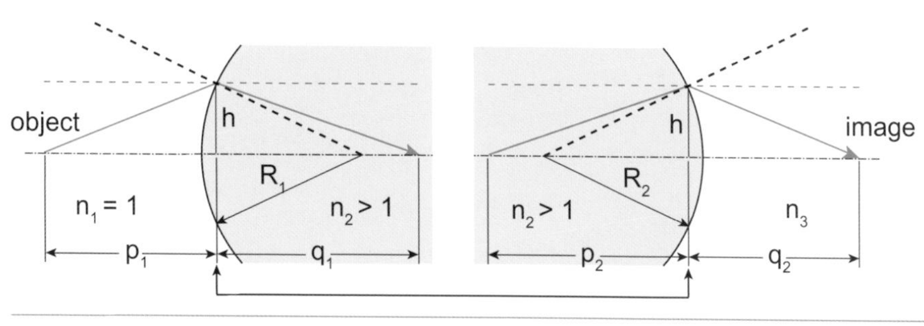

FIGURE C.1. The two surfaces of a simple lens.

The equation for the thin lens is then obtained by adding these equations with $n_3 = n_1$. With $p_1 = \infty$, we have $q_2 = f$, and thus we obtain the lensmaker's equation (Equation (7.3)):

$$\frac{1}{f} = \left(\frac{n_2}{n_1} - 1\right)\left(\frac{1}{R_1} - \frac{1}{R_2}\right). \quad \text{(C.2)}$$

Note that this equation is accurate only in the paraxial limit, where light rays make small angles with the optical axis and the lens is thin.

It is also interesting to consider the case where $n_1 \neq n_3$. This time we find that

$$\left(\frac{n_2 - n_1}{R_1} + \frac{n_3 - n_2}{R_2}\right) = \frac{n_1}{p_1} + \frac{n_3}{q_2}. \quad \text{(C.3)}$$

Setting $p_1 = \infty$ establishes that the left-hand side of Equation (C.3) is equal to n_3/f_3, where f_3 is defined to be the focal length in the region where the refractive index is n_3. Therefore, we obtain a new conjugate equation,

$$\frac{n_1}{p_1} + \frac{n_3}{q_2} = \frac{n_3}{f_3}. \quad \text{(C.4)}$$

This equation, for example, applies to the human eye, where the volume between the lens and the retina contains vitreous humor and $n_3 = 1.337$. By reversing the direction of the rays and setting $q_2 = \infty$, it is easy to show that $n_1/f_1 = n_3/f_3$.

Gaussian Optics and the Principal Planes

The essence of Gauss's theory can be captured without complicated equations. Using the ray in Figure 9.4(b) as an example, the argument goes as follows. The ray on the left side of surface A_1 is described by its distance (x_1) from the axis and the angle (α_1) it makes with the axis. This description is denoted by the symbol $\mathbf{V}_1$.

$$\mathbf{V}_1 = \begin{pmatrix} x_1 \\ \alpha_1 \end{pmatrix}.$$

The effect of surface A_1 on the ray is described by the "operator" $\mathbf{R}_1$. On the left side of surface A_1, the ray is described by $\mathbf{V}_1$ and on the right side it is described by $\mathbf{R}_1\mathbf{V}_1$ which we call $\mathbf{V}_1'$. This is an important step, because I have now introduced the concept of an "operator." An *operator* acts on the quantity to its right and produces a transformation. In short, I will just say that we multiply by $\mathbf{R}_1$ from the left. This is a powerful idea, because it can be extended to all the necessary steps of refraction and translation. Each refraction step has an operator $\mathbf{R}$ that depends on the power of the surface, and each translation step has an operator $\mathbf{T}$ that depends on the angle of the ray and the distance traversed.

Continuing the example, the ray incident on surface A_2 is described by $\mathbf{V}_2 = \mathbf{T}_{12}\mathbf{R}_1\mathbf{V}_1 = \mathbf{T}_{12}\mathbf{V}'_1$ and the ray exiting A_2 is described by $\mathbf{V}'_2 = \mathbf{R}_2\mathbf{T}_{12}\mathbf{R}_1\mathbf{V}_1$. All we are doing is multiplying from the left with the appropriate operator each time the ray passes through a surface or traverses a space. So far the operations only describe a single lens, and when D_{12} (in Figure 9.4) is small, it is a thin lens. At this point, all of the operations can be combined (multiplied) to obtain a single operator $\mathbf{M}_{12} = \mathbf{R}_2\mathbf{T}_{12}\mathbf{R}_1$.

Now suppose that a compound lens contains n elements. The same procedure is followed with a new **R** for each surface and **T** for each translation and eventually leads to the operator $\mathbf{M}_{1n}$. The operators $\mathbf{M}_{12}$ and $\mathbf{M}_{1n}$ that describe the effect of the single thin lens and the compound lens, respectively, are not identical, but they can be put into the same form by introducing translations (spaces) at the beginning and the end of the sequence for the compound lens. In the same notation, this gives a new operator $\mathbf{M}_{pp'} = \mathbf{T}'\mathbf{M}_{1n}\mathbf{T}$, where $\mathbf{M}_{pp'}$ has the proper form only when the translation operators are based on the proper principal planes, P and P'. Some readers will recognize **V** as a 2×1 matrix (column vector) and that operators **R** and **T** can be represented by 2×2 matrices.

A Macro Lens with Attachments

The working distances and magnifications were measured for a Sigma 105 mm macro lens with various combinations of extension tubes, diopters, and a tele-extender. The results are plotted as magnification versus working distance in Figure E.1. For each combination, measurements were made at far focus (infinity) and closest focus. The results are indicated with dots connected by red dotted lines. The points on the connecting lines have no significance and only serve to connect the endpoints. Each line is labeled to show the presence or absence of extension tubes (12 mm to 68 mm), diopters, and a Kenko 1.4X teleconverter. The red star shows the bare lens at close focus. Unconnected dots also refer to close focus.

The results show that a teleconverter in combination with extension tubes gives high magnification and large working distances. In each case, the extension tube was attached to the lens first. In other words the teleconverter was mounted directly on the camera body. The reversed 50 mm lens (20 diopter) gives high magnification, but at the cost of working distance. In Figure E.1 I show the measured distances (see Figure 12.3) and associated magnifications for all the combinations of the macro lens and attachments.

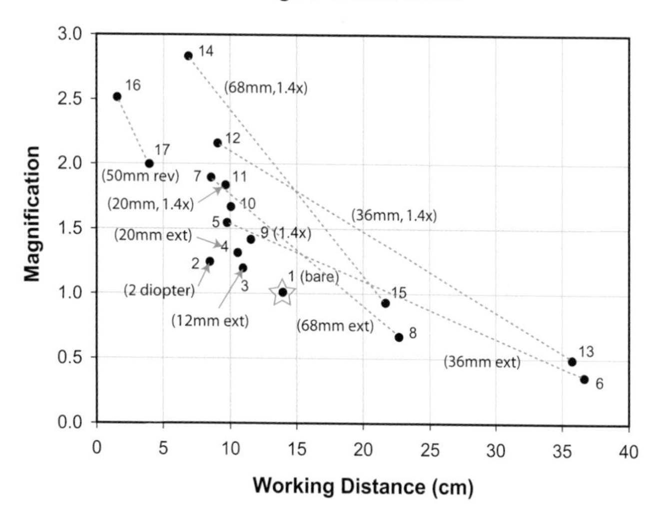

FIGURE E.1. Characterization of a 105 mm macro lens.

TABLE E.1.
The effects of diopters, extension tubes, and teleconverters

105 mm Lens+	Object to Sensor (cm)	Working Distance	Magnification
bare	31.4	14	1.00607
2 diopter-close	29.2	8.5	1.24282
12 mm ext-close	31.4	11	1.19347
20 mm ext-close	31.7	10.6	1.31188
36 mm ext-close	32.6	9.8	1.54367
36 mm ext-far	54.4	36.7	0.35265
68 mm ext-close	34.6	8.6	1.89219
68 mm ext-far	43.6	22.7	0.6658
1.4X Kenko-close	33.2	11.6	1.41494
12 mm+1.4x close	33.1	10.1	1.66552
20 mm+1.4x close	33.4	9.7	1.83273
36 mm+1.4x close	34.4	9.1	2.1554
36 mm+1.4x far	56	35.8	0.49109
68 mm+1.4x close	35.5	6.9	2.82811
68 mm+1.4x far	45.1	21.7	0.92794
50 mm rev close	26	1.6	2.51036
50 mm rev far	23	4	1.9927

APPENDIX F

Capturing Photons with Photographic Film

The discovery of photography in the early 19th century was essentially the discovery that certain silver salts, e.g., silver chloride (AgCl), silver bromide (AgBr) and silver iodide (AgI), are affected by light and can be processed to amplify the effect and make it permanent. These silver salts, known as silver halides, form crystals that visibly darken when exposed to bright light, and the process continues so that the crystals get darker and darker. This darkening effect is neither very sensitive nor stable and is not very useful for recording images. The enabling discovery for photography was that chemical processing, after exposure, can convert some of the exposed crystals into grains of metallic silver and also stop the darkening process. Indeed, there was a mystery. Even a low level of exposure does something to the crystals so that they can be "developed" later to obtain a silver grain record of the exposure. After exposure, the halide crystals contain a latent image which lasts until the crystals are developed.

Today, after many decades of experimentation, technological advances, and (more recently) scientific understanding, we have black-and-white and color films that provide stable, high-resolution images. However, they are still based on latent images stored in silver halide crystals. Modern film consists of a sturdy base about 100 μm to 200 μm thick, usually cellulose acetate, that serves to support multiple layers of emulsion held together with gelatin. For example, Fujicolor PROVIA 100 color-reversal film contains more than a dozen layers starting with a protective layer on top and ending with an antihalation layer to prevent reflection of light back into the emulsion. The thickness of the emulsion ranges from about 10 μm to 20 μm.

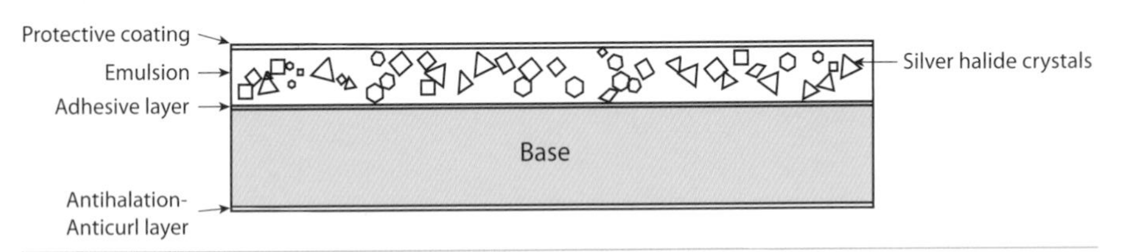

FIGURE F.1. Monochrome (black-and-white) film. The thickness is not to scale.

Since the photon capture step in film is always the same, it is easier to begin the discussion with black-and-white film as shown in Figure F.1. The silver halide crystals are roughly one micron in diameter and are randomly distributed.

In the commonly used silver bromide crystals, there is a lattice of silver ions (Ag^+) and bromide ions (Br^-) as shown in Figure F.2. An energetic photon can knock an electron out of a Br^- ion, and the electron has some probability of being captured by a lattice defect (shallow trap). If a mobile silver ion happens to be present in the same trap or reach the trap while the electron is still there, it will be reduced to a silver atom. This is not sufficient to contribute to the latent image because a single trapped atom is not stable. However, the newly minted atom will continue to hop from trap to trap, and it may encounter one or more other atoms to form a stable microcluster. The important point is that a certain level of exposure produces a proportionate number of stable microclusters of silver atoms in different silver bromide crystals.

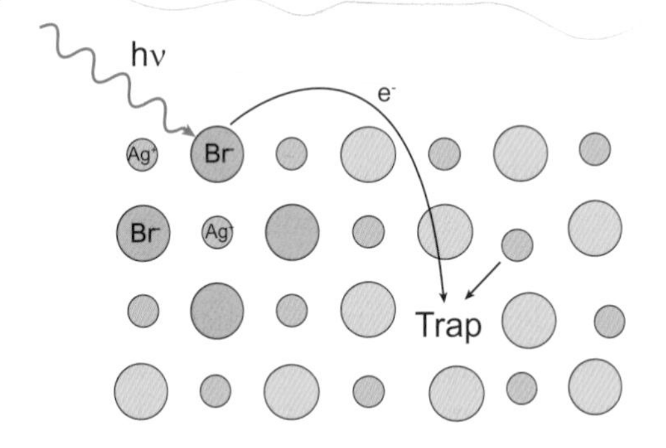

FIGURE F.2. Photo excitation of a silver bromide crystal.

The microclusters of silver atoms are called *sensitivity centers* or *development centers*, and they constitute the latent image. This latent image can be "developed out" by applying a developer solution with a reduction potential sufficient to reduce only those crystals containing microclusters to silver particles. In chemical terminology, *reduction* means adding an electron so that the silver ion (+1) is converted into metallic silver (0). There are usually two additional steps in film processing. First there is an acid "stop" bath (usually acetic acid) to terminate development, and then a "fixer" or "hypo" solution (usually sodium thiosulfate) to dissolve and wash away "unsensitized" crystals while leaving the silver particles (grains) in place. In this way, a few silver atoms in a cluster can act as a trigger for the conversion of billions of silver ions to metallic silver. The silver particles remain in the emulsion and constitute a negative image, since it is darker where there was more exposure. In going from latent image to silver image we effectively have a powerful photon multiplier.

As discussed in Chapter 13 (Note 1), silver halides are primarily sensitive to blue and UV radiation. Modern panchromatic film is possible because of sensitizers that are added to the emulsion. The sensitizers are dye molecules that are designed to absorb a certain color of light and to transfer electrons to a crystal grain. Sensitizers also make possible the construction of color film that can produce color negatives and positives (color-reversal film). Color perception and representation are discussed in detail in Chapter 15. Here it suffices to say that the "colors" of light can be represented with combinations of red (R), green (G), and blue (B) light in an additive process. The three colors that are necessary for use as pigments, or in any other subtractive process, are cyan, magenta, and yellow. This means that color film must start with the capture of red, blue, and green light and end up after development with cyan, magenta, and yellow dyes for positive or reversal film and the corresponding negative colors for print film.

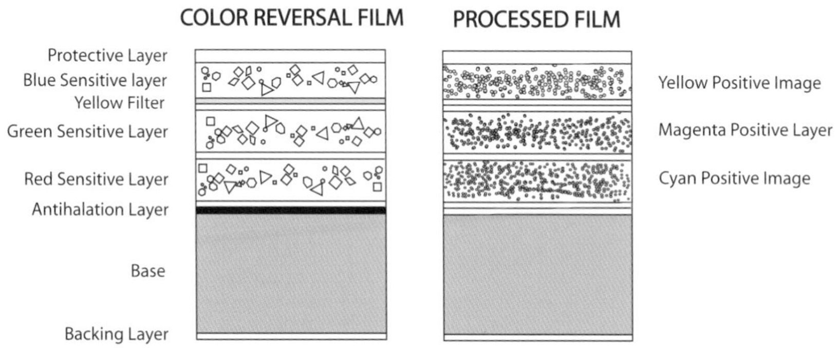

COLOR REVERSAL FILM
PROCESSED FILM

Protective Layer
Blue Sensitive layer
Yellow Filter
Green Sensitive Layer
Red Sensitive Layer
Antihalation Layer
Base
Backing Layer

Yellow Positive Image
Magenta Positive Layer
Cyan Positive Image

FIGURE F.3. Simplified cross-section of color-reversal film.

Figure F.3 presents a simplified picture of the emulsion layers in a film such as Fujichrome Provia. Each of the light-sensitive layers contains silver halide crystals along with appropriate sensitizers and dye-coupler molecules. All of the layers are sensitive to blue light, so the first layer is assigned to blue and is followed by a yellow filter to absorb the remaining blue light. The second and third layers are assigned to green and red light, respectively, and contain the required sensitizers. When the film is exposed to light, the light produces latent images in the three layers with concentrations of sensitized sites that are proportional to the exposure by the three colors of light. Thus the light is a stimulus that is represented in the film by latent images for R, G, and B colors.

The development process consists of the reduction of silver ions in sensitized crystals to metallic silver and the oxidization of the developer molecules, i.e., the removal of electrons. The oxidized developer molecules are then free to react with dye-coupler molecules to produce the required dyes in each layer. Films with this design are very practical now, but initially there was much difficulty keeping the dyes in their proper layers. For that reason Kodachrome, the first successful subtraction-process film, was quite different. In order to avoid the wandering dyes, Kodachrome was designed without any coupling-dyes in the film. It was essentially a multilayer black-and-white film with appropriate sensitizers, and the dyes were all added during processing. The result was an extremely complicated processing procedure with 28 steps and requiring large, specialized equipment and highly trained technicians. In addition to providing outstanding color and fine grain, the Kodachrome system had two major advantages: first, all the couplers were removed during processing, thereby improving stability, and, second, the layers in the emulsion were much thinner than those required to hold dye-couplers in competitive products, thus improving resolution.

Other successful schemes for manipulating dyes in color images were based on bleaching dyes already in place. A good example is Cibachrome (now Ilfochrome), that was engineered by Ciga-Geigy Corp. for color prints. In this material, stable azo dyes are incorporated into the light-sensitive layers. The development process selectively destroys dye molecules in the exposed areas. This results in vivid colors and archival prints. Cibachrome/Ilfochrome prints are among the best archival color prints that can be obtained.

The evolution of color photography extended over about a hundred years. There was a flurry of activity in the early 20th century motivated largely by the movie industry, and by the 1940s a number of practical color films were available to photographers. In 1947 Polaroid Corp. introduced the instant camera with self-developing film, and in 1963 the color instant camera was announced. This was perhaps the most sophisticated application of organic chemistry at the time. It also demanded reliable and accurate mechanical designs to ensure smooth movement of the print and delivery of the developer as soon as an exposure was made. This heroic achievement served its purpose for decades, but now has been swept away by the digital revolution. The Polaroid Corporation ceased to exist (except in name) in 2001, and other film companies are regularly announcing cancellations of products and developing services.

Historical note (holography, music, and color photography). Early attempts to make color photographs were all based on the additive color process, in which black-and-white negatives of the same scene were obtained through red, blue, and green filters. Positive images were made from the negatives and then projected through the corresponding color filters to recreate the color scene. This method, first demonstrated by James Clerk Maxwell (1861), continued to be used with a variety of ingenious variations well into the 20th century. The positive color method, at its best, was used by Sergey Prokudin-Gorsky to document the Russian Empire, 1905–1915. A single-plate version, in which the color filters took the form of colored grains of starch, was known as Autochrome. It produced a type of color pointillism that is still popular with collectors.

Pure color without dyes. In 1908, Gabriel Lippmann received the Nobel Prize in physics for color photography. Furthermore, he obtained true colors without any dyes. Lippmann was an important physicist and inventor, and he held positions as Professor of Experimental and Theoretical Physics at The Faculty of Sciences Laboratory in Paris and Director of the Research Laboratory (later part of the Sorbonne). He was a giant in his time but is relatively unknown today.

So what did Lippmann do to impress the Nobel committee? He proposed and demonstrated a way to encode color information into photographic plates so that colored images could be observed through the selective reflection of light. This is the same effect that gives rise to the iridescent color of hummingbirds through the diffraction of light. Diffraction was explained in Appendix B and shown in Figure B.4. Lippmann's idea was sophisticated and simple and almost impossible to realize with the materials available to him. He proposed an arrangement consisting of a plate with a thin, fine-grained panchromatic photographic emulsion in contact with a mercury mirror, as shown in Figure F.4. Light would be directed through the plate and emulsion and would be reflected back by the mercury surface. The phase of the reflected light is shifted by 180°, and the incoming and outgoing beams interfere to give nodes and loops (bright spots) separated by the wavelength of the light divided by 2. If we take red light from a He–Ne laser with $\lambda = 633$ nm as an example, we immediately see Lippmann's dilemma. The separation of the layers (lamina) in the latent image is 316 nm, and can only be recorded with extremely fine-grain film. If blue light is added, a set of layers with smaller separation is produced, and so on as other colors are added.

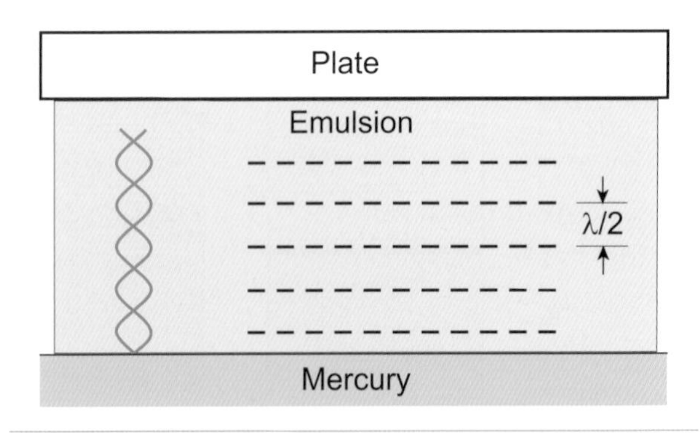

FIGURE F.4. Lippmann color photography illustrated with a single color of light.

Lippmann devoted considerable time to producing fine-grain, high resolution film and optimizing developer solutions; and he was able to obtain color images in spite of the absence of true panchromatic film. This feat can now be reproduced in an undergraduate laboratory by using commercial holographic plates with a resolution of more than 1000 lines/mm.

Lippmann's method of interference color photography was not practical for photographers, but his ideas have found new life in laser-based optical technology. Dennis Gabor was fascinated by Lippmann's ideas and set about making interference patterns in three-dimensional space. The result was the hologram, another Nobel Prize-winning invention. Reflection-type holograms can be made by exposing an emulsion to laser light scattered from an object and simultaneously to light directly from the laser at a different angle.

The reflecting planes in the developed film are similar to Lippmann's color photographs, and the reflection hologram is known as a *Lippmann hologram*. Also, Lippmann-Bragg volume holograms are used as filters in optical instruments, and the concept has been exploited for storing information in standing-wave memory devices where layers of information are recorded at different depths in an optically sensitive material. Gabriel Lippmann provides another example of an inventor who was ahead of his time. Fortunately, Lippmann was acknowledged and celebrated during his lifetime for a number of contributions.

Modern color film: Color film as we know it is based on the subtractive process. In the developed film, light is filtered through yellow, cyan, and magenta layers before it reaches the eye, as shown in Figure F.3. A surprisingly complete description of this method with the correct order of sensitized layers and the concept of coupler-dyes was described by Rudolph Fischer in 1911. Fischer was prescient, but the time was not right for the realization of his dream. His sensitizers and dyes diffused between layers, and he could not obtain satisfactory color photographs. It would be almost 30 years before the technical problems could be overcome and transparency film could be manufactured with dyes that would stay put.

The magnificent (three-color) Kodachrome film for movies and 35 mm stills was brought to the market by Kodak in 1936 and not retired until 2009. This product was the brainchild of brilliant amateurs. Leopold Godowsky, Jr., and Leopold Mannes, who shared interest in music and photography, met as boys in New York City. They recognized the mediocre quality of so-called color movies before 1920, and they decided to set up a laboratory to experiment with color photography. They also worked as musicians, a vocation that would continue throughout their lives. Godowsky studied violin and chemistry and physics at UCLA and later mathematics at Columbia. Meanwhile, Mannes studied physics and musicology at Harvard.

Fortunately, Godowsky and Mannes were able to continue their collaboration by mail, and in 1922 they established a real laboratory in New York. Soon they were on the track that would lead to Kodachrome, and at the same time they were becoming successful musicians. Through a chance encounter with Lewis Strauss, who was a junior associate with Kuhn, Loeb and Co., they were able to obtain financial backing. Their work was so impressive that in 1930 they were invited to move to Rochester to work in the well-equipped Kodak Laboratories. The result of this collaboration was the development of a highly successful color-subtractive film that proved to have amazing archival properties, and their coworkers quipped that Kodachrome was created by "God and Man." In 2005, Godowsky and Mannes were inducted into the National Inventors Hall of Fame.

Further Reading

S. J. Williamson and H. Z. Cummins. *Light and Color in Nature*. New York: Wiley, 1983.

M. R. Peres (Editor). *Focal Encyclopedia of Photography, Fourth Edition*. Amsterdam: Elsevier, 2007.

N. A. Rosenblum. *World History of Photography, Third Edition*. New York: Abbeville Press, 1997.

S. Matsugasaki. "Lippmann Color Photography." *J. Optics* 22 (1991), 267–273.

R. E. Schwall and P. D. Zimmerman. "Lippmann Color Photography for the Undergraduate Laboratory." *Am. J. Phys.* 38 (1970), 1345–1349.

APPENDIX G

Microelectronics and the Path to Digital Sensors

A detailed timeline of developments in photography and digital image-processing can be found in the *Focal Encyclopedia of Photography, 4th Edition*. Here I discuss a few of the more spectacular advances. It is difficult to pinpoint the beginning of "electronic photography." Photographs were "scanned" and transmitted over wire in the 19th century, and there were digitized images (Bartlane System) and radiophotograph fax machines in the 1920s. Xerography appeared in 1938, and Russell Kirsch scanned a photograph of his baby son into a computer in 1957. (*Life* honored Kirsch's image in 2003 as one of the "100 photographs that changed the world.")

The most important advance for digital photography, however, was the invention of the transistor in 1948 (Bardeen, Brattain, and Shockley; Nobel Prize, 1956). This single step replaced the vacuum tube with a solid-state device and initiated a revolution in electronics. A simple form of the vacuum tube (triode) contained a heated filament, a cathode, a grid, and an anode. It looked a bit like a light bulb, and its function was to control the current between the cathode and the anode with a small voltage on the grid. It acted like a valve, or gate. The transistor, its replacement, also controls the flow of current by means of a small voltage (or current), but it is smaller, faster, more reliable, and uses less energy. The transistor was probably the most important invention of the 20th century.

The next major step in the quest for smaller, more-efficient electrical circuits was the design of the integrated circuit in 1958 by Jack Kilby (Nobel Prize, 2000) and Robert Noyce. Kilby was a new employee with Texas Instruments (TI), who was working on the problem of dealing with large numbers of components in electrical circuits. His seminal idea was to put the entire circuit in a single piece of semiconductor (a chip). The time was right, and he was able to demonstrate a prototype in a few months. Kilby remained at TI, and he obtained over 60 patents, including the patent for the electronic calculator. Another integrated-circuit design was developed independently by

Noyce at Fairchild Semiconductor a few months later. The design was not only different, but it also solved some significant problems in Kilby's design. Noyce, who had been a co-founder of Fairchild, went on to co-found Intel Corporation.

With the development of the transistor and integrated circuits, the race for higher-and-higher densities of circuit components was on. This is one of the great success stories of technology in the 20th century. In 1965, Gordon Moore, another founder of Intel, noted that the number of transistors that could be put into an integrated circuit was increasing exponentially with time. In fact, the number of circuit elements continued to double every couple of years from 1971 through at least 2004, and other measures of the microelectronic revolution, such as hard-disk storage capacity, RAM capacity of computers, and pixels per unit area in sensors were also increasing exponentially. This trend is referred to as Moore's Law.

In the 1960s, the stage was set for the development of image sensors. The breakthrough occurred in 1969, when George E. Smith and Willard S. Boyle at AT&T Bell Labs invented charge-coupled devices (CCDs). These were initially just memory devices with shift registers that could shift charges from cell to cell without loss. However, it was soon realized that the charges could originate via the photoelectric effect in photo cells. Images were being captured with CCD image sensors in 1970, and digital photography was born. Smith and Boyle have received numerous awards for their invention, and they were both elected members of the National Academy of Engineering. In addition to recognition by the microelectronics community, they were honored with the Progress Metal of the Photographic Society of America, and in 2009 they shared the Noble Prize in Physics. The other half of the 2009 award went to Charles K. Kao for the development of ultrapure fiber optic cables that have revolutionized communications.

This is not quite the end of the story for image sensors, since it is evident that complementary metal-oxide semiconductor (CMOS) devices are appearing in many high-end cameras in place of CCDs. CMOS detectors are active-pixel sensors that contain an array of pixels, each of which has transistors and a readout circuit in addition to a photodetector. (These descended from primitive MOS sensors from the 1960s and early active-pixel MOS sensors invented by Peter J. W. Nobel (1968) and Savvas G. Chamberlain (1969).) These sensors are easily fabricated by standard semiconductor techniques and are, therefore, cheaper than CCDs, but initially they were viewed as second-rate sensors. However, research at the Jet Propulsion Laboratory in the 1990s and at camera companies since that time has changed the picture. CMOS sensors are now recognized as competitive with CCDs in image quality in addition to offering advantages in cost and energy requirements.

The culmination of this story for our purposes is, of course, the invention of the digital camera. This achievement is credited to Steven J. Sasson. Sasson was a young electrical engineer working for Eastman Kodak Company in 1975. His assignment was to determine whether a camera could be built using solid-state electronics instead of film (here the term camera implies image capture and recording in a portable unit). The result was an 8-pound, 0.01-megapixel prototype, which was patented in 1978 by Sasson and his then-supervisor Gareth Lloyd. It was only in 2009, after digital imaging became a multi-billion-dollar industry, that Sasson started receiving adequate recognition. He has received an "Innovation Award" from the *Economist* magazine and an honorary degree from the University of Rochester.

Irradiance and Illuminance Units

Flux is the measure of how much of something is flowing through unit area in unit time. Energy per square meter per second is an example. This can be measured in units of joules/meter^2 sec or J/(m^{2}s). Since J/s is defined as a watt (W), we can write the flux as W/m^2. So far I have been describing the intensity of light, which is the same as irradiance. That is great for a physicist, but for engineers dealing with illumination, it is not adequate.

The simplest way to proceed is to note that illumination engineers measure illuminance in units of lumens per square meter or lm/m^2, and this flux is defined as a lux (lx) (see ["Lux" 09]). Therefore, the lumen, under certain conditions, is proportional to energy per second, and it looks a bit like a watt. The difference between watts and lumens is that a watt of green light appears to be much brighter than a watt of red light, while lumens have been designed so that a lumen of any color in the visible region appears to be equally bright. This sounds like psychophysics, and it is. We have to ask human observers to match the brightness of different colors of light, or to measure by some means the sensitivity of the average human eye to light of different frequencies. This has, in fact, been done for light-adapted (photopic region) and dark-adapted (mesopic region) human eyes, and the associated luminosity functions are shown in Figure H.1.

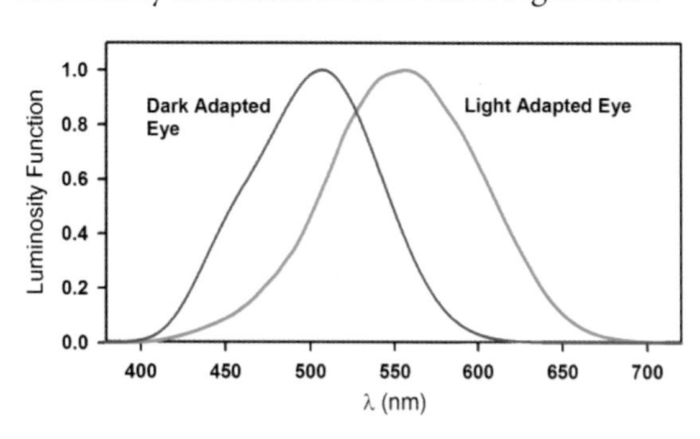

FIGURE H.1. Luminosity functions for the human eye: dark-adapted (blue curve) and light-adapted (green curve).

The standard lm is approximately 1/683 W or 1 W produces 682 lm at 555 nm, and values for the other wavelengths are based on the photopic luminosity curve. For example, at 400 nm, 1 W produces 0.27 lm, and at 700 nm, 1 W produces 2.8 lm.

Suppose a light beam containing a variety of colors (wavelengths) is viewed by a human observer. Furthermore, let the spectral-power density be given by the function J(λ) in watts/(unit wavelength). The apparent brightness of this beam can be computed if the standard luminosity function $\overline{y}(\lambda)$ is known:

$$\text{Flux}(\text{lumens}) = 683\,\frac{\text{lm}}{\text{W}}\int \overline{y}(\lambda) J(\lambda)\, d\lambda\,,$$

where the range of integration is 400 nm to 700 nm. This just means that each increment of the power curve has to be adjusted by the luminosity curve before it is added to obtain the total number of lumens.

This comparison of units is necessary for understanding the characteristic curves for film and sensors. However, I have not related the lumen to the other units in common use. This is tedious, but necessary for understanding the literature. Consider a light source at the center of a sphere having unit radius. If the source emits 1 lm/steradian uniformly in all directions the output is defined as 1 candela (cd). (A steradian is defined as the solid angle subtended by 1 m^2 on the surface of a sphere having a radius of 1 m.) Furthermore, 1 footcandle (fc) is defined as 1 lm/ft^2 just as we defined 1 lux as 1 lm/m^2.

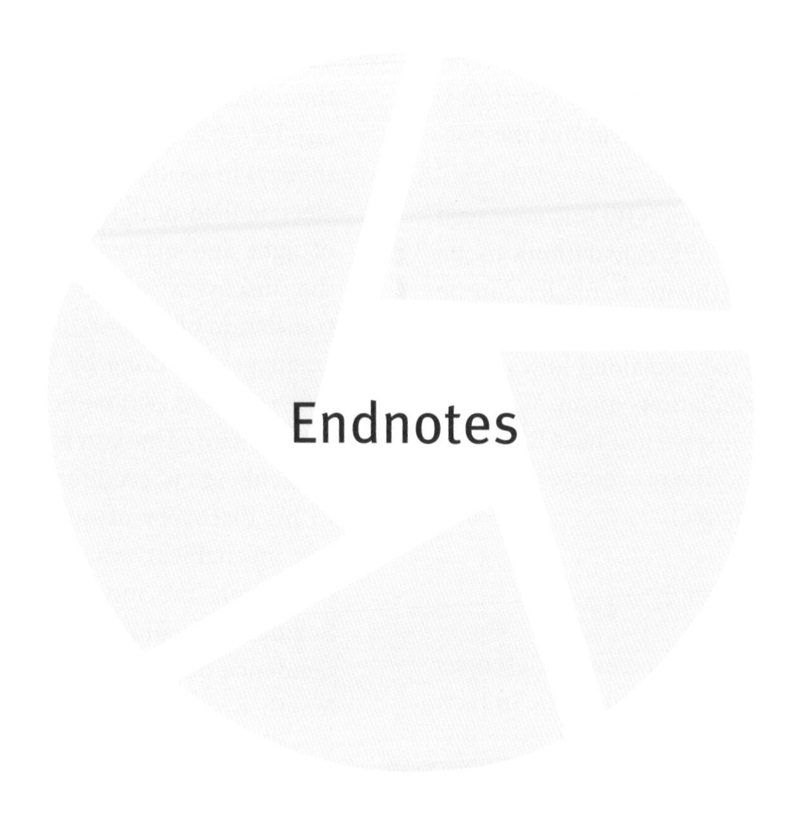

Endnotes

Chapter 3

1. Many Kodak film types have been discontinued. For example, Kodak removed from production 828 (28 mm × 40 mm) in 1985, 127 (1 ⅝ in. × 2 ½ in.) and 620 (2 ¼ in. × 3 ¼ in.) in 1995, and Kodachrome slide film in 2009. A few professionals and serious amateurs still use 4 in. × 5 in. sheet film. The most recent film introduction was number 240, Advanced Photo System (APS), in 1996. This film, which provides a transparent magnetic recording medium for image information in addition to the traditional silver halide film, did not succeed in displacing 35 mm as the most popular amateur film.

Chapter 6

1. According to the theory of relativity, the speed of light is a constant in the universe, so how can light have a reduced speed in matter, e.g., water or glass? In fact, it turns out that the index of refraction can be less than 1 and even less than 0. So the refractive index is much more subtle than one might think. Feynman's interpretation of the index is as follows: The speed of light always has the value c. When light exits matter, however, the electric field that is detected results from all the electrical charges present. When light passes through matter the electrons in all the atoms are disturbed and in turn contribute to the radiation field. The field we measure after the interaction of light with matter is the sum of all fields, and the resulting superposition exhibits a phase (location of nodes) that is approximately the same as that calculated by assuming that the speed of light in matter is given by c/n. The only thing that matters in optics is the phase of the wave, not how long it took to pass through the glass.

2. In the terminology of the calculus, the derivative of P(ABC) with respect to x is equal to 0 at the minimum, and this determines the corresponding value of x.
3. Quantum electrodynamics (QED) is very well developed and it permits calculations to be made with precision. The mathematics can be daunting, but Nobel Prize-winning physicist Richard P. Feynman has explained how easy it is to see what the calculations mean. We don't have to understand why, we just have to accept the way nature works. The procedure is ridiculously simple (in principle).

Chapter 7

1. The development of lenses and even telescopes is credited to craftsmen, not scientists. In fact, philosophers dismissed lenses since their operation was not consistent with existing (European) theories of vision that were based on the emission of rays from the eye. They concluded that spectacles could only be a disturbing factor, and that one should not trust things seen through lenses. The emission theory of vision had, in fact, been disproved by Ibn al-Haytham in the 11th century by consideration of after-images in the eye and the pain that results from looking at the sun.
2. Amazingly, about 50% of US college students still believed in visual emissions in 2002. See [Winer et al. 02]; as this paper reports, roughly one half of college students believe that vision depends on emanations from the eyes, with traditional education failing to overcome this belief.

Chapter 8

1. Wikipedia is the gateway to most of the terms used here. For the most part, Wikipedia is reliable for noncontroversial scientific information, but it should be used with a critical eye. When in doubt, read other sources.
2. The Sellmeier equation was given a reasonable theoretical foundation by the Dutch physicist Hendrik Lorentz in 1892 (Drude-Lorentz theory) in terms of oscillating electrons in the electric field of light rays. Of course, our view of light and matter changed completely with the discovery of quantum mechanics. The translation of Lorentz's equation into quantum language was done by John van Vleck (1925) and by Hans Kramers and Werner Heisenberg (1926). The latter paper, co-authored by Heisenberg, is credited with leading directly to his discovery of quantum mechanics (matrix mechanics), which was published shortly thereafter. The modern interpretation of the Sellmeier equation is as follows: The major dependence of the index function $n(\lambda)$ on wavelength λ results from absorption of light at the wavelengths $\lambda_1 = \sqrt{C_1}$, $\lambda_2 = \sqrt{C}\text{pt}_2$, and $\lambda_3 = \sqrt{C_3}$. The contributions to $n(\lambda)$ of the absorptions at λ_1, λ_2, and λ_3 are determined by the coefficients B_1, B_2, and B_3, respectively. These coefficients are related to the strengths of the absorptions and can, in principle, be computed by the methods of quantum mechanics. Each absorption is associated with the transition of an electron from its ground energy state to a higher-energy state. For example, in the transition at λ_1, a photon of light with energy $h\nu_1$ is absorbed by an electron to produce the energy change $\Delta E = h\nu_1 = E_1^{\text{excited}} - E_1^{\text{ground}}$.
3. The lensmaker's equation shows that, for a given lens, the quantity $f \cdot (n - 1)$ is a constant. The application of differential calculus immediately relates the fractional change in focal length f to the change in refractive index n: $\delta f/f = -\delta n / (n - 1)$. Here δf and δn represent the changes in the focal length and the refractive index, respectively. The right-hand side of this equation is known as the dispersive power of the material, and the Abbe number is just an approximation of the inverse of this quantity.

Chapter 9

1. See Chapter 3 of [Klein 70], a standard college textbook; Chapter 10 of [Fowles 89], an accessible text that treats the thick lens; and Chapter 4 of [Born and Wolf 65], which is heavy reading but is authoritative.

Chapter 12

1. Plus diopters have little effect on the F-number. It is easy to understand what is happening here. As shown in Figure 12.4, neither the distance to the sensor nor the aperture are changed by the added diopter. Therefore, the effective F-number, which is the distance (q) to the sensor divided by the aperture (δ), is unchanged. Another way to understand the effect of the added diopter is as follows. The diopter effectively reduces from f_{Prime} to f_{Combo} as described in Equation (12.6), and this means that the defined F-number is reduced to $N \cdot f_{\text{Combo}}/f_{\text{Prime}}$. At the same time, the magnification becomes m = $f_{\text{Prime}}/f_{\text{Diopter}}$, which is equal to $(f_{\text{Prime}}/f_{\text{Combo}}) - 1$. The final step is to use the expression for effective F-number to obtain $N_{\text{eff}} = N \cdot (f_{\text{Combo}}/f_{\text{Prime}}) \cdot (1 + m) = N$. This more lengthy derivation shows that the intensity increase expected from the reduced F-number is cancelled exactly by the increase in magnification. Furthermore, the magnification results from the fact that the lens to sensor distance (q) exceeds f_{Combo}. This amounts to "extension," which is discussed in Section 12.3.

Chapter 13

1. Filters have a long history in photography. The earliest photographic emulsions were primarily sensitive to blue and UV light, but in the 1870s the German photochemist and photographer Hermann Vogel learned how to extend the sensitivity into the green region by adding dyes to silver bromide. The result was the orthochromatic plate, which still lacked sensitivity to red light. Adolph Miethe, an astrophotographer who succeeded Vogel as Professor of Photography at the Royal Technical University in Berlin, was able to extend sensitivity to the complete visible spectrum in 1903 to produce panchromatic emulsions through the addition of the sensitizing dye, ethyl red. There was still excessive sensitivity in the blue region, however, and realistic tones could only be achieved by photographing through a yellow filter.

   Photographic filters that selectively attenuate colors were invented by Frederick Wratten for dealing with the limited tonal ranges of available emulsions. The ability to modify tones in black-and-white images became very important to photographers, and in 1912 Eastman Kodak purchased Wratten's company and continued the production of "Wratten" filters. Later, after color emulsions became available, filters were even more important. Each color emulsion was designed with a particular white balance. For example, film for daylight use might be balanced for 5500 K light. Such an emulsion would produce a decidedly yellow-orange cast if used with tungsten incandescent light (3400 K). The way around the problem was either to purchase a film balanced for 3400 K light or to use a blue filter (Wratten 80B). Conversely, "tungsten" film balanced for 3400 K could be used in daylight by adding an amber filter (Wratten 85).

   The introduction of digital cameras along with computer post processing of images has drastically reduced the need for colored filters. These cameras have built-in UV and IR filters to avoid false colors, and the white balance can be set in the camera or (in some cases) postponed for later processing. Overall attenuation to permit long exposures is still desirable, and selection of the direction of polarization remains an important option. In addition, ultraviolet (UV) and infrared (IR) photography require that visible light be blocked with a filter.

2. Unfortunately, Bouguer did not get full credit for the absorption law because he did not provide a compact mathematical expression to describe the effect. Johann Lambert's contribution was to apply the calculus to sum up the contributions of infinitesimally thin layers to show that the attenuation decreases exponentially as described by the equation $I/I_0 = 10^{-OD}$, where OD is defined to be optical density. [Actually, Lambert's expression was closer to $I = I_0 \exp(-a \cdot l)$, where a is a constant that denotes the absorption coefficient of the material and l is the thickness of the sample. There is a simple conversion between the forms of the equation, given by $OD = a \cdot l/\ln(10)$.] So how did August Beer get into the act? His contribution was to relate the OD, or *absorbance*, as chemists call it, to the concentration of particles or molecules in a transparent liquid solvent or perhaps the atmosphere. Thus, in place of Lambert's constant a, Beer introduced the product of a molar absorption coefficient ε and the concentration of particles. This turned out to be very important for chemists because, given the molar absorption coefficient, the concentration of a solution could be determined from a measurement of the optical density (or absorbance).

3. Some forms of crystalline tourmaline are naturally dichoric, so that one crystal will polarize light and a second crystal can be rotated to block the polarized light completely. Tourmaline absorbs light too strongly to be used as a practical polarizer, however. Crystals of herapathite (quinine sulphate per-iodide) were also found to be dichroic in the 19th century, but it turned out to be very difficult to grow the crystals large enough to make useful polarizers. In spite of that, Carl Zeiss was able to market Bernotar single-crystal herapathite polarizing filters for photography. Information about the crystal-growing techniques they used was apparently lost during World War II.

   Commodity polarizers date from the 1920s, when the scientist and inventor Edwin Land demonstrated that a suspension of tiny crystals (ground in a mill) aligned by electric or magnetic fields can also polarize light. With the insight that aligned crystallites or even aligned molecules can substitute for a single crystal, he went on to develop polarizing sheets of PVA (polyvinyl alcohol) doped with iodine atoms. The trick was that the sheets were stretched in the manufacturing process in such a way that the polymer molecules were aligned and the attached iodine atoms formed chains. The chains of atoms act like short segments of metallic wire that strongly absorb light polarized parallel to their length. It is the electrons, free to move in the direction of the atomic chain, that interact with the electric field With this technique and various improvements, Land's company, Polaroid Corporation, was able to produce efficient and inexpensive polarizers for sunglasses, photography, and a variety of other uses.

4. Suppose that polarized light with the maximum amplitude **E** and the intensity I (Figure 13.11) is incident on a linear polarizer so that the component $\mathbf{E}_1$ is transmitted and $\mathbf{E}_2$ is blocked. The transmitted light has amplitude $\mathbf{E}_1 = \mathbf{E} \cdot \cos\theta$ and intensity $I_1 = I\cos^2\theta$. If the incident light is unpolarized, angle θ is distributed over 360°, or 2π radians, and one must obtain an average of $\cos^2\theta$ in order to compute the intensity of the transmitted light. The result for the transmitted intensity is $I_{\text{polarized}} = I/2$. This explains why 50% of unpolarized light is transmitted by a linear polarizer.

   The expression for circularly polarized light is more interesting. At any instant in time, say time 0, the amplitude varies according to position along the z-axis (see Figures 13.8–13.10). For the x-component, the amplitude is described by $\mathbf{E}_1 \cos(k \cdot z)$, where $k = 2\pi/\lambda$. If the y-component is represented by $\mathbf{E}_2 \cos(k \cdot z)$, with $\mathbf{E}_1 = \mathbf{E}_2$, then the result is just a linearly polarized wave at 45° as shown in Figure 13.11. The introduction of a quarter-wave plate, however, changes things so that the y-component becomes $\mathbf{E}_2 \sin (k \cdot z)$ and the resultant rotates around the z-axis. Note that a quarter-wave plate requires no filter factor because the electric field of the light beam is not attenuated. In fact, a second quarter-wave plate (rotated 90°) can be used to restore the original linear polarization without loss. The x- and y-components shown here were used in the program Mathcad to generate the sine curves in Figures 13.9 and 13.10 and the helix in Figure 13.10.

Chapter 14

1. Visual acuity (VA) can defined by the ability to resolve alternating black and white lines. For example, VA = $1/x$, where x is the size of a line pair or the black line spacing measured in arc minutes. Standard acuity (VA = 1) is then defined as the ability to resolve 1 min of arc, and the notation 20/20 indicates that the observer can resolve detail at 20 ft. that the standard observer resolves at 20 ft. It is claimed that the maximum human acuity is something less than 20/10 or VA = 2.0. It should be noted that VA is strongly affected by the level of illumination, and the highest levels are achieved when the light level is greater than 0.1 Lambert.

Chapter 17

1. The Gaussian probability distribution function is defined to be

$$G(x) = 1/(\sigma\sqrt{2\pi})\exp\left[(-x-\mu)^2/(2\sigma^2)\right]$$

where the center is at $x = \mu$ and σ is the standard deviation. The half-width at half-height of this distribution is given by $\sigma\sqrt{2\ln 2}$, and 68% of the area enclosed by the curve lies within a distance of σ of the center. See http://en.wikipedia.org/wiki/Normal_distribution.

2. The Fourier transform of a lineshape function resolves the function into sinusoidal components and provides the amplitude and phase (shift) of each component. The transformation is reversible so that the sine and cosine curves can be added together to reproduce the original curve. The MTF displays the amplitudes only, and phase information is lost. Phase distortions may be severe when aberrations are present, but the MTF is still an important indicator of lens quality. For illustrations of frequency components see [Palmer 99, p. 159].

3. See [Jones 58]. This paper presents the relationships among the PSF, LSF, and the MTF. The PSF function contains information about resolution and contrast in all directions, while the LSF defines only one direction. A good discussion can be found in [Smith 97, Chapter 25]. Hardcopies and a free electronic version are available through the website, http://www.dspguide.com/editions.htm.

Bibliography

[Atkins 09a] B. Atkins. "MTF and SQF, Part 4: Subjective Quality Factor." Available at http://www.bobatkins.com/photography/technical/mtf/mtf4.html, 2009.

[Atkins 09b] B. Atkins. "MTF and SQF, Part 5: SQF and Popular Photography." Available at http://www.bobatkins.com/photography/technical/mtf/mtf5.html, 2009.

[Bach 09] M. Bach. "84 Optical Illusions and Visual Phenomena." Available at http://www.michaelbach.de/ot/, 2009.

[Balter 00] M. Balter. "Paintings in Italian Cave May Be Oldest Yet." *Science* 290 (2000), 419–421.

[Balter 08] M. Balter. "Going Deeper into the Grotte Chauvet." *Science* 321 (2008), 904–905.

[Born and Wolf 65] M. Born and E. Wolf. *Principles of Optics.* Oxford: Pergamon Press, 1965.

[Bourke 09] P. Bourke. "Computer-Generated Angular Fisheye Projections." Available at http://local.wasp.uwa.edu.au/~pbourke/miscellaneous/domefisheye/fisheye/, 2009.

[Bowmaker and Dardnall 80] J. K. Bowmaker and H. J. A. Dardnall. "Visual Pigments of Rods and Cones in a Human Retina." *J. Physiol.* 298 (1980), 501–511.

[Campbell and Robson 68] F. W. Campbell and J. G. Robson. "Application of Fourier Analysis to the Visibility of Gratings." *J. Physiol.*, 197 (1968), 551–566.

[Canon 03] Canon, Inc. Lens Group. *EF Lens Work III: The Eyes of EOS.* Canon, Inc., 2003.

[Carl Zeiss 05] Carl Zeiss AG. "In Memory of Ernst Abbe." *Innovation* 15 (2005), 4–7. Available at http://www.zeiss.com/c12567a100537ab9/Contents-Frame/616f09d817f19f75c125692000402cc8.

["CIE" 09] "CIE 1931 Color Space." *Wikipedia.* Available at http://en.wikipedia.org/wiki/CIE_1931_color_space, 2009.

[CIPA 09] CIPA Standardization Committee. "Sensitivity of Digital Cameras (Published July 27, 2004)." CIPA DC-004-Translation-2004. Technical report, Camera & Imaging Products

Association. Available at http://www.cipa.jp/english/hyoujunka/kikaku/pdf/DC-004_EN.pdf, 2009.

[Clark 09a] R. N. Clark. "Notes on the Resolution and Other Details of the Human Eye." Available at http://www.clarkvision.com/imagedetail/eye-resolution.html, 2009.

[Clark 09b] R. N. Clark. "Digital Imaging Information." Available at http://www.clarkvision.com/imagedetail/evaluation-canon-s70/index.html, 2009.

[Covington 07] M. A. Covington. *Digital SLR Astrophotography*. Cambridge: Cambridge University Press, 2007.

[Crane 64] E. M. Crane. "An Objective Method for Rating Picture Sharpness: SMT Acutance." *Jour. SMPTE,* 73 (1964), 643–647.

["Daniel Schwenter" 09] "Daniel Schwenter." *Wikipedia*. Available at http://en.wikipedia.org/wiki/Daniel_Schwenter, 2009.

[Dersch 09] H. Dersch. "Panorama Tools." Available at http://www.path.unimelb.edu.au/~dersch/, 2009.

[De Valois et al. 82] R. L. De Valois, D. G. Albrecht, and L. G. Thorell. "Spatial Frequency Selectivity of Cells in Macaque Visual Cortex." *Vision Res.,* 22 (1982), 545–559.

[Duck 88] M. J. Duck. "Newton and Goethe on Color: Physical and Physiological Considerations." *Annals of Science* 45 (1988), 507–519.

[Efros 09] A. Efros. "Computational Photography." Carnegie Mellon University; http://graphics.cs.cmu.edu/courses/15-463/, 2009.

[Fauvel et al. 88] J. Fauvel, R. J. Wilson, M. Shortland, and R. Flood (Editors). *Let Newton Be!* New York: Oxford University Press, 1988.

[Fowles 89] G. R. Fowles. *Introduction to Modern Optics, Second Edition*. New York: Dover, 1989.

[Fraser et al. 05] B. Fraser, C. Murphy, and F. Bunting. *Color Management, Second Edition*. Berkeley, CA: Peachpit Press, 2005.

[Gazzaniga 08] M. S. Gazzaniga. *Human: The Science Behind What Makes Us Unique*. New York: HarperCollins, 2008.

[Granger 74] E. M. Granger. "Subjective Assessment and Specification of Color Image Quality." *S.P.I.E.,* 46 (1974), 86–92.

[Granger and Cupery 72] E. M. Granger and K. N. Cupery. "An Optical Merit Function, which Correlates with Subjective Image Judgements." *Photogr. Sci. Eng.,* 16 (1972), 221–230.

[Giorgianni and Madden 98] E. J. Giorgianni and T. E. Madden. *Digital Color Management, Encoding Solutions.* Reading, MA: Addison-Wesley, 1998.

[Hattar et al. 02] S. Hattar, H.-W. Liao, M. Takao, D. M. Berson, and K.-W. Yau. "Transduction by Retina Ganglion Cells That Set the Circadian Clock." *Science* 295 (2002), 1065–1070.

[Hayes 08] B. Hayes. "Computational Photography." *American Scientist* 96 (2008), 94–98.

[Heavens 91] O. S. Heavens. *Optical Properties of Thin Films.* New York: Dover, 1991.

[Howe and Purves 05] C. Q. Howe and D. Purves. "The Müller-Lyer Illusion Explained by the Statistics of Image-Source Relationships." *Proc. Natl. Acad. Sci. USA* 102 (2005), 1234–1239.

[Howe et al. 05] C. Q. Howe, Z. Yang, and D. Purves. The Poggendorff Illusion Explained by Natural Scene Geometry. *Proc. Natl. Acad. Sci. USA* 102 (2005), 7707–7712.

[Hubel 95] D. H. Hubel. *Eye, Brain, and Vision,* Scientific American Library, 22. New York: W. H. Freeman, 1995. Available online at http://hubel.med.harvard.edu/bcontex.htm.

[Imatest 09] Imatest. "Introduction to SQF." Available at http://www.imatest.com/docs/sqf.html, 2009.

[Jacobson 09] D. Jacobson. "Lens Tutorial." Available at http://photo.net/learn/optics/lensTutorial, 2009.

[Jones 58] R. C. Jones. "On the Point and Line Spread Functions of Photographic Images." *J. Opt. Soc. Am.* 48 (1958), 934–937.

[Juza 09] E. A. Juza. "My Sigma 360mm f/7.1 Macro (June 11, 2007)." Available at http://www.juzaforum.com/forum-en/viewtopic.php?f=3&t=41, 2009.

[Kingslake 89] R. Kingslake. *A History of the Photographic Lens*. New York: Academic Press, 1989.

[Klein 70] M. V. Klein. *Optics*, New York: Wiley, 1970.

[Koch and Tononi 08] C. Koch and T. Tononi. "Can Machines Be Conscious?" *IEEE Spectrum* 45 (2008), 55–59.

[Kolb et al. 09] H. Kolb, E. Fernandez, and R. Nelson. "Visual Acuity." Webvision: The Organization of the Retina and Visual System. Available at http://webvision.med.utah.edu/KallSpatial.html, 2009.

[Koren 09] N. Koren. "Monitor Calibration and Gamma." Available at http://www.normankoren.com/makingfineprints1A.html, 2009.

[Land 83] E. H. Land. "Recent Advances in Retinex Theory and Some Implications for Cortical Computations: Color Vision and the Natural Image." *Proc. Natl. Acad. Sci. US* 80 (1983), 5163–5169.

[Lee 99] T. S. Lee. "Image Representation Using 2D Gabor Wavelets." *IEEE Trans. Pattern Analysis and Machine Intelligence,* 18 (1996), 959–970.S.

[Littlefield 09] R. Littlefield. "Theory of the 'No-Parallax' Point in Panorama Photography." Available at http://www.janrik.net/PanoPostings/NoParallaxPoint/TheoryOfTheNoParallaxPoint.pdf, 2009.

[Livingstone 02] M. Livingstone. *Vision and Art: The Biology of Seeing*. New York: H. N. Abrams, Inc., 2002.

["Lux" 09] "Lux." *Wikipedia*. Available at http://en.wikipedia.org/wiki/Lux, 2009.

["MacAdam Ellipse" 09] "MacAdam Ellipse." *Wikipedia*. http://en.wikipedia.org/wiki/MacAdam_ellipse, 2009.

[Macleod 99] A. Macleod. "The Early Days of Optical Coatings." *J. Opt. A: Pure Appl. Opt.* 1 (1999), 779–783.

[Melles Griot 09a] Melles Griot. Optics Guide. Available at http://www.mellesgriot.com/products/optics/toc.htm, 2009.

[Melles Griot 09b] Melles Griot. Glossary. Available at http://www.mellesgriot.com/glossary/wordlist/glossarylist.asp, 2009.

[Meuris 04] J. Meuris. *René Magritte*. Los Angeles: TASCHEN America, 2004.

[Miyamoto 64] K. Miyamoto. "Fish Eye Lens." *J. Opt. Soc. Am.* 54 (1964), 1060–1061.

[Nave 09] C. R. Nave. "Dispersion." HyperPhysics. Available at http://hyperphysics.phy-astr.gsu.edu/hbase/geoopt/dispersion.html, 2009.

[Newton 52] I. Newton. *Opticks; or, A treatise of the reflections, refractions, inflections & colours of light.* New York, Dover, 1952. Based on the 4th edition, London, 1730.

["Optical Illusion" 09] "Optical Illusion." *Wikipedia.* Available at http://en.wikipedia.org/wiki/Optical_illusion, 2009.

[Ostroff 84] E. Ostroff. "Photographic Enlarging: A History." *Photographic Science and Engineering* 28 (1984), 54–89.

[Palmer 99] S. E. Palmer. *Vision Science.* MIT Press, Cambridge, MA, 1999.

[Pinker 99] S. Pinker. "How Much Art Can a Brain Take?" *The Independent* February 7, 1999. Adapted from *How the Mind Works,* London: Penguin, 1999.

[Ribe and Steinle 02] N. Ribe and F. Steinle. "Exploratory Experimentation: Goethe, Land, and Color Theory," *Physics Today* 55:7 (2002), 3–49.

[Rovamo et al. 98] J. Rovamo, H. Kukkonen, and J. Mustonen. "Foveal Optical Modulation Transfer Function of the Human Eye at Various Pupil Sizes," *J. Opt. Soc. Am. A* 15 (1998), 2504–2513.

[Sacks 95] O. Sacks. "The Case of the Colorblind Painter." In *An Anthropologist on Mars.* New York: Vintage Books, 1995.

[Sepper 88] D. L. Sepper. *Goethe Contra Newton* Cambridge: Cambridge University Press, 1988.

[Smith 97] S. W. Smith. *The Scientist and Engineer's Guide to Digital Signal Processing.* San Diego: California Technical Publishing, 1997.

[Smith et al. 04] D. R. Smith, J. B. Pendry, and M. C. K. Wiltshire. "Metamaterials and Negative Refractive Index." *Science* 305 (2004), 788–792.

[Smolyaninov et al. 07] I. I. Smolyaninov, Y.-J. Hung, and C. C. Davis. "Magnifying Superlens in the Visible Frequency Range." *Science* 315 (2007), 1699–1701.

["Solid Angle" 09] "Solid Angle." *Wikipedia*. Available at http://en.wikipedia.org/wiki/Solid_angle, 2009.

[Solso 94] R. L. Solso. *Cognition and the Visual Arts*. Cambridge, MA: MIT Press, 1994.

[Solso 03] R. L. Solso. *The Psychology of Art and the Foundation of the Conscious Brain*. Cambridge, MA: MIT Press, 2003.

[Spring et al. 09] K. R. Spring, T. J. Fellers, and M. W. Davidson. "Introduction to Charge-Coupled Devices (CCDs)." Available at http://www.microscopyu.com/articles/digitalimaging/ccdintro.html, 2009.

["Standard Illuminant" 09] "Standard Illuminant." *Wikipedia*. Available at http://en.wikipedia.org/wiki/Standard_illuminant, 2009.

[Turchetta et al. 09] R. Turchetta, K. R. Spring, and M. W. Davidson. "Introduction to CMOS Image Sensors." Available at http://micro.magnet.fsu.edu/primer/digitalimaging/cmosimagesensors.html, 2009.

[van Walree 09] P. van Walree. "Photographic optics with illustrations of Seidel aberrations." Available at http://www.vanwalree.com/optics.html, 2009.

["Vignetting" 09] "Vignetting." *Wikipedia*. Available at http://en.wikipedia.org/wiki/Vignetting#Pixel_vignetting, 2009.

[Werblin and Roska 07] F. S. Werblin and B. Roska. "The Movies in Our Eyes." *Scientific American* 296 (2007), 72–79.

[Westfall 93] R. S. Westfall. *The Life of Isaac Newton*. Cambridge: Cambridge University Press, 1993.

[Winer et al. 02] G. A. Winer, J. E. Cottrell, V. Gregg, J. S. Fournier, and L. A. Bica. "Fundamentally Misunderstanding Visual Perception: Adults' Belief in Visual Emissions." *Amer. Psych.* 57:6–7 (2002), 417–424.

[Wood 11] R. W. Wood. *Physical Optics*. New York: Macmillan Co., 1911.

[Zakia 02] R. D. Zakia. *Perception and Imaging, Second Edition*. Boston: Focal Press, 2002.

[Zeki 99] S. Zeki. *Inner Vision: An Exploration of Art and the Brain*. New York: Oxford University Press, 1999.

Index